**Progress in Nonlinear Differential Equations and Their Applications**
Volume 36

Martin Flucher

# Variational Problems with Concentration

Springer Basel AG

Martin Flucher
SHERPA'X AG
Glutz-Blotzheimstr. 1
4503 Solothurn
Switzerland

1991 Mathematics Subject Classification 35J50; 35R35, 35J60, 35Q99

A CIP catalogue record for this book is available from the Library of Congress, Washington D.C., USA

Deutsche Bibliothek Cataloging-in-Publication Data

**Flucher, Martin:**
Variational problems with concentration / Martin Flucher. -
Basel ; Boston ; Berlin : Birkhäuser, 1999
(Progress in nonlinear differential equations and their applications ; Vol. 36)
ISBN 978-3-0348-9729-7 ISBN 978-3-0348-8687-1 (eBook)
DOI 10.1007/978-3-0348-8687-1


Originally published by Birkhäuser Verlag, Basel, Switzerland in 1999
Softcover reprint of the hardcover 1st edition 1999
Printed on acid-free paper produced of chlorine-free pulp. TCF ∞

ISBN 978-3-0348-9729-7

9 8 7 6 5 4 3 2 1

*To my teacher Jürgen Moser*

# Contents

# Chapter 1

# Introduction

To start with we describe two applications of the theory to be developed in this monograph: Bernoulli's free-boundary problem and the plasma problem.

### Bernoulli's free-boundary problem

This problem arises in electrostatics, fluid dynamics, optimal insulation, and electro chemistry. In electrostatic terms the task is to design an annular condenser consisting of a prescribed conducting surface $\partial\Omega$ and an unknown conductor $A$ such that the electric field $\nabla u$ is constant in magnitude on the surface $\partial A$ of the second conductor (Figure 1.1). This leads to the following *free-boundary problem* for the electric potential $u$.

$$\begin{aligned} -\Delta u &= 0 \quad \text{in } \Omega \setminus A, \\ u &= 0 \quad \text{on } \partial\Omega, \\ u &= 1 \quad \text{on } \partial A, \\ \frac{\partial u}{\partial \nu} &= Q \quad \text{on } \partial A. \end{aligned}$$

The unknowns are the free boundary $\partial A$ and the potential $u$. In optimal insulation problems the domain $\Omega \setminus A$ represents the insulation layer. Given the exterior boundary $\partial\Omega$ the problem is to design an insulating layer $\Omega \setminus A$ of given volume which minimizes the heat or current leakage from $A$ to the environment $\mathbb{R}^n \setminus \Omega$. The heat leakage per unit time is the capacity of the set $A$ with respect to $\Omega$. Thus we seek to minimize the capacity among all sets $A \subset \Omega$ of equal volume. The corresponding Euler Lagrange equation is Bernoulli's free-boundary problem with the constant $Q$ playing the role of a Lagrange multiplier. In order to transform this problem into a more standard variational problem we introduce the volume integrand

$$F(t) := \begin{cases} 0 & (t < 1), \\ 1 & (t \geq 1). \end{cases}$$

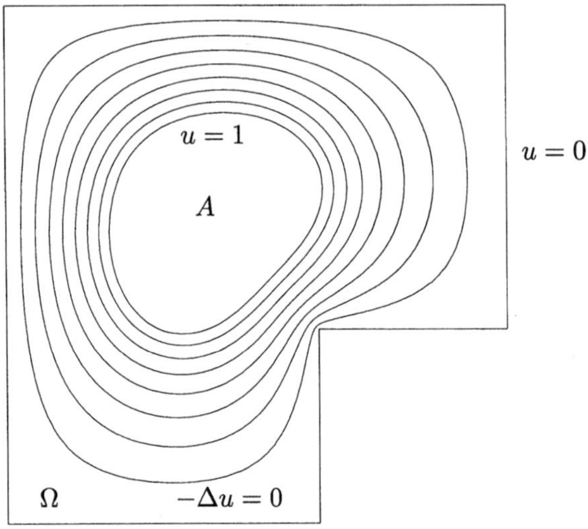


Figure 1.1: Solution of Bernoulli's free-boundary problem with equipotential lines.

Solutions of Bernoulli's free-boundary problem are then constructed by maximizing

$$\int_{\Omega} F(u) \quad = \quad |\{u \geq 1\}|$$

in a suitable space of functions with zero boundary data and fixed Dirichlet integral $\int_{\Omega} |\nabla u|^2$. The optimal insulation layer is finally obtained as a sublevel set of the optimal potential $\Omega \setminus A = \{0 < u < 1\}$. In terms of this example one of the aims of the present monograph is to analyze the geometrical properties of a small conductor $A$. We will characterize both its shape and location in a form that is suitable for engineering purposes. In particular several numerical schemes for the numerical computation of its asymptotic location will be devised (Section 8.4). A detailed discussion of Bernoulli's free-boundary problem is given in Chapter 14.

**Plasma problem**

Our second application arises in plasma physics (Section 3.4). The aim is to determine the shape of a plasma ring in a tokamak. In its simplest version the problem can be described as follows. Given the cross-section $\Omega$ of the tokomak, find the plasma region $A \subset \Omega$ and the magnetic potential $u$ such that:

$$\begin{aligned} -\Delta u &= \lambda(u-1)_+^{q-1} \text{ in } \Omega \setminus \partial A, \\ u &= 0 \text{ on } \partial\Omega, \\ u &= 1 \text{ on } \partial A, \end{aligned}$$

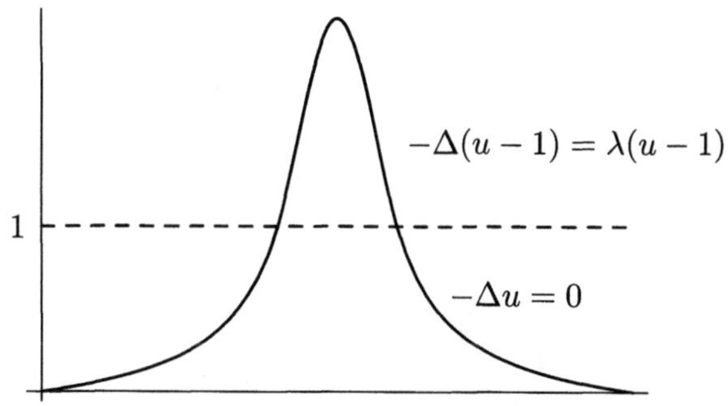


Figure 1.2: Solution of plasma problem on ball in $\mathbb{R}^3$ with $q=2$.

$$u \quad \text{is} \quad C^1 \quad \text{across} \quad \partial A.$$

Figure 1.2 shows an example with $q = 2$. Variational solutions of this free-boundary problem can be obtained by maximizing

$$\int_\Omega F(u)$$

with

$$F(t) \quad = \quad \begin{cases} (t-1)^q & (t \geq 1), \\ 0 & (t < 1) \end{cases}$$

among functions with zero boundary data and fixed Dirichlet integral. The resulting Euler equation is the above free-boundary problem involving a Lagrange multiplier $\lambda$. Again we are interested in the shape and location of thin plasma rings.

**General boundary value problem**

More generally we consider nonlinear Dirichlet problems of the form

$$\begin{aligned} -\mu \operatorname{div}\left(|\nabla u|^{p-2}\,\nabla u\right) &= f(u) \text{ in } \Omega, \\ u &= 0 \text{ on } \partial\Omega \end{aligned} \tag{1.1}$$

involving the *p-Laplacian* in the range $1 < p \leq n$ where $n$ denotes the space dimension. Important special cases are $p = 2$ corresponding to the *semilinear equation*

$$-\mu\,\Delta u \quad = \quad f(u)$$

and the *conformal case* $p = n$. As to the *viscosity parameter* $\mu$ we consider the *vanishing viscosity limit* $\mu \to 0$.

### Variational formulation

We have chosen to solve the boundary value problem (1.1) via the *constraint variational problem*

$$\sup\left\{\int_\Omega F(u)\ :\ \int_\Omega |\nabla u|^p \le \varepsilon^p,\ u=0 \ \text{ on } \ \partial\Omega\right\} \tag{1.2}$$

with $F' = f$ and $F(0) = 0$. We focus on the properties of *low energy solutions* $u_\varepsilon$ with small $\varepsilon$. Equation (1.1) is the *Euler Lagrange equation* corresponding to the constraint variational problem. The viscosity parameter plays the role of a Lagrange multiplier. Equivalently we could use the more popular *mountain pass method.* With this method solutions of (1.1) are obtained as the critical points of the functional

$$u \quad\mapsto\quad \frac{\mu}{p}\int_\Omega |\nabla u|^p - \int_\Omega F(u).$$

With the variational approach the p-Laplacian case is conceptually similar to the semilinear case. On the contrary, with PDE methods the passage from the linear to the nonlinear problem is difficult.

### Low energy limit and principal goals

Our principal goal is to understand the asymptotic behaviour of the solutions of (1.2) in the *low energy limit* $\varepsilon \to 0$. In particular we want to:

1. Show that low energy solutions concentrate at a single point, i.e. that they form a spike near the concentration point and are small elsewhere.

2. Describe the shape of the spike.

3. Identify the concentration point in geometrical terms.

These results are new in the literature except for special cases. Once they are established we are in position to provide a precise description of the positive solutions of the boundary value problem (1.1) with small viscosity.

In some applications the unknown function $u$ represents the temperature and $\mu$ the heat conductivity. If so a maximum point of $u$ is called a *hot spot.* As the heat conductivity tends to zero the hot spots accumulate at some point. This is the PDE characterization of the concentration point. Instead we will characterize it as the point where the Dirichlet energy concentrates. This is the variational point of view.

### Hypotheses and function spaces

We consider the boundary value problem (1.1) for nonlinearities satisfying

$$f'(0) \quad = \quad 0.$$

This is the *zero mass case.* The natural function space for solutions of (1.2) in this case is the *Sobolev space* $D^{1,p}(\Omega)$. It can be defined as the closure of the space of all smooth functions with compact support with respect to the norm

$$\|\nabla v\|_p \quad := \quad \left(\int_\Omega |\nabla v|^p\right)^{1/p} .$$

The variational approach restricts us to the range $1 < p \leq n$. The domain $\Omega$ can be a bounded or unbounded subdomain of $\mathbb{R}^n$ satisfying the exterior ball condition.

In contrast to the zero mass case the *positive mass case* $f'(0) < 0$ is well studied in the literature. The corresponding semilinear equations are of the form

$$-\Delta u + cu \quad = \quad g(u)$$

with $c = -f'(0) > 0$ and $g'(0) = 0$. The appropriate function space for this case is $H_0^{1,p}(\Omega)$. On domains of finite volume the spaces $D^{1,p}$ and $H_0^{1,p}$ coincide. However, in general $D^{1,p}$ is not contained in $L^p$. In the remaining case $f'(0) > 0$ the supremum in (1.2) is infinite. In view of these facts we assume that the integrand satisfies the growth condition

$$(1.3) \qquad 0 \;\leq\; F(t) \;\leq\; \begin{cases} c\,|t|^{p^*} & (p < n), \\ c\left(\exp\left(\alpha\,|t|^{\frac{n}{n-1}}\right) - 1\right) & (p = n). \end{cases}$$

Here

$$p^* \quad := \quad \frac{np}{n-p}$$

denotes the *critical Sobolev exponent*; $c$ and $\alpha$ are positive constants. In addition $F$ is supposed to be upper semi continuous and $F \not\equiv 0$ in the $L^1$ sense. These conditions are obviously satisfied for the volume and the plasma integrand. Assuming the above growth condition (1.2) is always finite. In the *subconformal case* $p < n$ this follows from *Sobolev's inequality*

$$(1.4) \qquad \int_\Omega |v|^{p^*} \;\leq\; S^* \left(\int_\Omega |\nabla v|^p\right)^{p^*/p}$$

which holds for all functions in $D^{1,p}(\Omega)$. The best constant $S^*$ is independent of the domain. In the conformal case $p = n$ the Dirichlet integral $\int_\Omega |\nabla u|^n \, dx$ is invariant under conformal transformations of the domain. For this reason the analysis of conformal low energy limits differs substantially from that for the subconformal case. In particular the critical Sobolev exponent becomes infinite. However, $D^{1,n}$ is not contained in $L^\infty$. Instead it embeds into the Orlicz space given by the Trudinger-Moser inequality. According to this inequality the supremum (1.2) is finite for small $\varepsilon$.

The growth condition (1.3) is crucial for most results of this study. Even though it is not a serious restriction at infinity, it is one near zero. In particular it restricts us to the zero mass case. On the other hand the integrand $F$ is allowed to be discontinuous. Thus we can treat certain free-boundary problems as well as a large class of nonlinear eigenvalue problems.

### Generalized Sobolev inequality

Several properties of the critical power integrand $F(t) = |t|^{p^*}$ can be generalized to more general integrands satisfying the growth condition (1.3). They satisfy the generalized Sobolev inequality

$$\int_\Omega F(u) \quad \leq \quad F^* \|\nabla u\|_p^{p^*}.$$

This is the starting point of our analysis. The generalized Sobolev constant

$$F^* \quad := \quad \sup\left\{ \int_{\mathbb{R}^n} F(w) \; : \; w \in D^{1,p}(\mathbb{R}^n), \; \|\nabla w\|_p \leq 1 \right\} \tag{1.5}$$

is finite due to the growth condition (1.3). Like the standard Sobolev constant it is independent of the domain.

### Concentration compactness alternative

We will prove concentration of low energy extremals by means of a generalized concentration compactness alternative. It covers the alternative of P.L. Lions [91] for the critical power integrand as a special case. Roughly speaking a *concentration compactness alternative* (abbreviated as CCA) says that one of the following alternatives holds true.

A Concentration: After normalization the energy densities tend to a Dirac measure.

B Compactness: After normalization the functions under consideration converge in norm.

In our case *concentration* of a sequence $(v_\varepsilon)$ means that the Dirichlet densities

$$|\nabla v_\varepsilon|^p \, dx \quad \stackrel{*}{\rightharpoonup} \quad \delta_x \;\text{ as }\; \varepsilon \to 0$$

converge in the sense of measures to a Dirac measure concentrated at some point $x \in \overline{\Omega}$. To prove concentration of low energy extremals we consider the normalized solutions $v_\varepsilon = u_\varepsilon/\varepsilon$ of the variational problem (1.2). By the generalized concentration compactness alternative we are left to exclude compactness. It turns out that compactness can only occur if the domain is all of $\mathbb{R}^n$. Thus we conclude that in the zero mass case low energy solutions concentrate at a single point unless the domain is all of $\mathbb{R}^n$. This is the first main result. It describes the asymptotic behaviour of low energy extremals on a coarse scale.

### Compactness criterion

Every concentration compactness alternative leads to a *compactness criterion* of the following form.

> If the supremum of a functional is strictly bigger than the maximal limit that can be achieved along concentrating sequences then the supremum is attained.

An argument of this type first occurs in the pioneering paper of Aubin [10] on the Yamabe problem. In a more general form it has been popularized by Brézis and Nirenberg [27, 25]. Typically the supremum of a functional is given by a global Sobolev constant $S$ as in (1.5). The maximal limit of the functional along a sequence concentrating at some point $x$ is given by what we call a *local Sobolev constant* $S^{\mathrm{loc}}(x)$. The above general compactness criterion says that if

$$\sup_{x\in\overline{\Omega}} S^{\mathrm{loc}}(x) \quad < \quad S \tag{1.6}$$

then the variational problem for $S$ admits an extremal function. Otherwise it may happen that a maximizing sequence for $S$ concentrates. If it concentrates it can only concentrate at a point $x$ with $S^{\mathrm{loc}}(x) = S$. To apply this criterion we will compute the local Sobolev constant explicitly and estimate the global one from below by construction of comparison functions. Specific problems to be attacked by this compactness criterion are the question of existence or nonexistence of extremal functions for the best Sobolev constant with anisotropic Dirichlet energy, a similar question for conformal metrics, and the construction of extremal functions for the Trudinger-Moser inequality.

### Asymptotic shape of low energy extremals

Next we want to determine the shape of the spikes formed by low energy extremals. As the energy decreases the spikes become steeper at a specific rate. To undo the steepening we rescale the independent variable according to

$$w_\varepsilon(y) \quad = \quad u_\varepsilon\left(x_\varepsilon + \varepsilon^{\frac{p}{n-p}} y\right).$$

Using the compactness-splitting-vanishing alternative of P.L.Lions [89] we show that the rescaled functions $w_\varepsilon$ converge to an entire extremal. An entire extremal is a solution of the variational problem (1.2) with $\Omega = \mathbb{R}^n$. Thus it is desirable to gather more information on entire extremals. In contrast to the positive mass case little has been done in the zero mass case. We contribute the following second order decay estimate:

$$w(x) \quad = \quad c\,|x|^{p-n}\,(1 + o(|x|)) \quad \text{as} \quad |x| \to \infty.$$

Moreover we provide an integral representation for the constant c. Low energy solutions for the volume or the plasma integrand lead to small solutions of the respective free-boundary problems. By the above convergence result and the subsequent remark the corresponding free boundaries form a family of concentrating asymptotically spherical surfaces (Figure 14.5 on page 128).

As a byproduct of the decay estimate we solve a symmetry problem raised by P.L. Lions. He conjectured that for positive nonlinearities positive entire solutions are radially symmetric and strictly decreasing in radial direction. We give an affirmative answer to this conjecture. Hence the rescaled low energy extremals converge to a function that can be easily computed by solving a radial problem. The method of moving planes and its generalizations do not apply. Even in the simplest examples entire extremals are not smooth enough for this method to apply. Instead we use our decay estimate and a characterization of the functions for which Schwarz symmetrization preserves the Dirichlet integral. To complete the picture of entire extremals it would be desirable to prove uniqueness of radial ground states. However, this problem is unsolved.

### Robin functions

Before proceeding we need to introduce a geometrical concept. The following convention is used to distinguish the semilinear case $p = 2$ from the general case $1 < p \leq n$. A term with the prefix p- refers to the general case while no prefix is used in the semilinear case. The p-Green's function is defined as the solution of the singular Dirichlet problem

$$\begin{aligned} -\operatorname{div}\left(|\nabla G_x|^{p-2}\,\nabla G_x\right) &= \delta_x \text{ in } \Omega, \\ G_x &= 0 \text{ on } \partial\Omega. \end{aligned}$$

Separation of the radial kernel

$$K(r) = c r^{\frac{p-n}{p-1}}$$

yields what we call the p-Robin function

$$\tau(x) := \lim_{y \to x} K(|y-x|) - G_x(y).$$

A related geometrical quantity is the p-harmonic radius $\rho$ defined as $K(\rho(x)) = \tau(x)$. A minimum point of the p-Robin function or maximum point of the p-harmonic radius is called a p-harmonic center of $\Omega$. For bounded domains satisfying the exterior ball condition it is always an interior point.

### Concentration points

So far we know that low energy extremals concentrate at a single point. Locating these points in geometrical terms is our third goal. We will show that the concentration point is a p-harmonic center of the underlying domain. This assertion will be proven for certain classes of integrands. However, we expect it to hold for all nonlinearities satisfying the growth condition (1.3).

The strategy of the proof is as follows. We derive a second order expansion for $\int_\Omega F(u_\varepsilon)$ with respect to the energy $\varepsilon$. In the subconformal case the leading term is the generalized Sobolev constant which does not depend on the domain. Only the second term involves the geometry via the p-Robin function. Up to lower order terms we have

$$\int_\Omega F(u_\varepsilon) \quad \approx \quad \varepsilon^{p^*} F^* \left(1 - c\,\varepsilon^{\frac{p}{p-1}} \min_\Omega \tau\right)$$

with an explicit positive constant $c$ depending on the integrand and the dimension. The expansion on the right with the p-Robin function evaluated at the point of concentration is an upper bound for the left side. This follows from a second order expansion for the p-capacity of small balls:

$$\operatorname{pcap}_\Omega(B_x^r) \quad \approx \quad \operatorname{pcap}_{\mathbb{R}^n}(B_x^r) + (p-1)\operatorname{pcap}_{\mathbb{R}^n}(B_x^r)^{\frac{p}{p-1}}\tau(x).$$

For the lower bound we construct comparison functions concentrating at an arbitrary given point. This is achieved by means of p-harmonic transplantation (Chapter 10). Combining these estimates yields the approximation formula for $\int_\Omega F(u_\varepsilon)$. By optimality of the extremals $u_\varepsilon$ and positivity of the constant $c$ the value of the p-Robin function at the concentration point must be a minimum.

In the conformal case already the leading term depends on the value of the p-Robin function at the concentration point. In terms of the solutions $v_\varepsilon$ of the corresponding radial problem on the unit ball we have

$$\int_\Omega F(u_\varepsilon) \quad \approx \quad \max_\Omega \rho^n \int_{B_0^1} F(v_\varepsilon).$$

Thus the p-harmonic radius must be maximal at the concentration point. In both cases the dependence of the first nontrivial term on the p-Robin function or p-harmonic radius allows us to conclude that the concentration point is a p-harmonic center. We summarize the main results on low energy solutions mentioned so far.

> For each $\varepsilon > 0$ let $u_\varepsilon$ be a solution of the variational problem (1.2) with an integrand satisfying (1.3) on a domain satisfying the exterior ball condition. As $\varepsilon \to 0$ these solutions behave as follows.
>
> 1. They concentrate at a single point unless the domain is all of $\mathbb{R}^n$.
> 2. On a microscopic scale near the concentration point the spikes are asymptotic to an entire extremal.
> 3. The concentration point is a p-harmonic center of the domain.

**Renormalized energy**

In the applications of this monograph the p-Robin function plays the role of a *renormalized energy*. This concept has been introduced by Bethuel, Brézis, and

Hélein [21] in the context of Ginzburg-Landau vortices. In this application each vortex carries infinite energy. However, it is possible to divide the energy into two parts in a natural way. The first part is infinite and is independent of the geometry. The second part is finite and carries geometrical information. The latter part is the renormalized energy. It is a function of the vortex positions and depends on the geometry of the underlying domain. The vortices tend to arrange themselves such as to minimize the renormalized energy.

In Chapter 15 a similar program is carried out for point vortices in an incompressible planar fluid. The Green's function describes the coupling of vortex pairs while the Robin function describes the interaction of an individual vortex with the walls of the flow container. The vorticity $\omega_p$ of the vortex with position $z_p$ is conserved as long as no collision occurs. The renormalized kinetic energy

$$E(z_1, z_2, \ldots) \;:=\; \sum_{q>p} \omega_p \omega_q G_{z_q}(z_p) - \frac{1}{2} \sum_p \omega_p \tau(z_p)$$

is a Hamiltonian and therefore an integral of motion. The motion of the vortices is governed by the corresponding Hamiltonian equations. We contribute a result on long time existence of vortex pairs. It is optimal in the sense that three vortices can collide in finite time. The proof is based on a careful analysis of the boundary behaviour of the Robin function.

Returning to elliptic boundary value problems the analogy with the Ginzburg-Landau vortices arises in the vanishing viscosity limit. For higher order solutions of the Euler Lagrange equation also *multi point concentration* occurs. In this situation the renormalized energy is similar to that for fluid vortices with the vortex positions replaced by the concentration points. Bahri, Li, and Rey [14] considered positive solutions of the singularly perturbed Dirichlet problem

$$-\Delta u \;=\; u^{\frac{n+2}{n-2}-\varepsilon}.$$

Supposing multi point concentration they have shown that the set of concentration points is critical for the renormalized energy. Single point concentration for this special problem is discussed in Section 13.4.

The relevance of the critical points of the Robin function for semilinear Dirichlet problems involving the critical Sobolev exponent was first pointed out by Schoen [117] and Bahri [13]. Detailed work in this direction has been carried out by Rey [107, 108, 109, 110, 111] and Egnell [40, 41, 42]. The new discovery of the present study is that concentration at the critical points of the Robin function is a more general phenomenon. In particular it is not restricted to integrands whose leading term is the critical power.

#### Related problems

Single point concentration occurs in a variety of variational problems. However, the location of the concentration points is sensitive to anisotropy and boundary conditions. We specify some problems of this type not covered in the present monograph.

a) As $\varepsilon \to 0$ the extremal functions of the variational problem

$$\sup\left\{\int_\Omega G(\cdot)F(u)\ :\ \int_\Omega |\nabla u|^p \le \varepsilon^p,\ u=0 \text{ on } \partial\Omega\right\}$$

concentrate at a maximum point of the function $G$. Results of this type for the Dirichlet problem

$$-\mu\,\Delta u \quad=\quad -V(\cdot)u + f(u)$$

have been obtained by del Pino and Felmer in [36, 37] and subsequent papers.

b) The extremal functions of

$$\sup\left\{\int_\Omega F(u)\ :\ \int_\Omega \nabla u\cdot A(\cdot)\nabla u \le \varepsilon^2,\ u=0 \text{ on } \partial\Omega\right\}$$

concentrate at a minimum point of the determinant of $A$ (cp. also Section 5.1).

c) The Dirichlet problem

$$-\mu\,\Delta u \quad=\quad -u + u^q$$

with subcritical exponent $q>1$ has been studied by Ni and Wei [101], and others. In the vanishing viscosity limit the hot spots of positive solutions accumulate at a point of maximal distance from the boundary. This is an example for the positive mass case.

d) The solutions of the corresponding homogeneous Neumann problem concentrate at a boundary point of maximal mean curvature, for instance at a corner of minimal opening angle. At such points high values of the integrand $\frac{1}{q}u^q - \frac{1}{2}u^2$ can be reached without paying too much for the Dirichlet integral. For the above equation this has been shown by Ni and Takagi [99, 100]. For

$$-\Delta u \quad=\quad u^{\frac{n+2}{n-2}} - \lambda u$$

with $\lambda \to \infty$ similar results have been obtained by Adimurthi, Pacella, Yadava [5] and Rey [112, 113].

**Acknowledgments**

This research has benefited from my previous work on semilinear problems with Catherine Bandle, Alain Brillard, Björn Gustafsson, Stefan Müller, Martin Rumpf, Michael Struwe, and Juncheng Wei. I want to thank all of them for fruitful and inspiring collaboration. I also thank Haïm Brézis, Patricio Felmer, Bernd Kawohl, Moshe Marcus, Yi Li, and Olivier Rey for calling my attention to recent contributions to the field. The figures 1.1, 8.2, 8.3, 8.4, 14.1, 14.3, 14.4, and 14.5 have been produced by Martin Rumpf using GRAPE. The remaining figures are MAPLE and MATHEMATICA plots.

# Chapter 2

# P-Capacity

The capacity is a key notion in the calculus of variations. In this chapter we collect some basic properties that will be used later. The *p-capacity* of a set $A \subset \Omega$ with respect to $\Omega$ is defined as

$$(2.1) \qquad \operatorname{pcap}_\Omega(A) \;:=\; \inf\left\{\int_\Omega |\nabla u|^p \,:\, u \in D^{1,p}(\Omega),\ u \geq 1 \text{ on } A\right\}.$$

By definition the p-capacity is monotone with respect to $A$ and $\Omega$. In analogy to the semilinear case we call

$$(2.2) \qquad \operatorname{pmod}_\Omega(A) \;:=\; \operatorname{pcap}_\Omega^{\frac{1}{1-p}}(A)$$

the *p-modulus* of $A$ with respect to $\Omega$. A function which realizes the infimum (2.1) is called a p-capacity potential. If the infimum is finite a unique p-capacity potential exists. It satisfies the Euler Lagrange equation

$$\begin{aligned} -\operatorname{div}\left(|\nabla u|^{p-2}\,\nabla u\right) &= 0 \text{ in } \Omega \setminus A, \\ u &= 0 \text{ on } \partial\Omega, \\ u &= 1 \text{ on } \overline{A}. \end{aligned}$$

Integration by parts leads to the boundary integral representation

$$(2.3) \qquad \operatorname{pcap}_\Omega(A) \;=\; \int_{\partial A} |\nabla u|^{p-1}.$$

Note that integrals over a domain in $\mathbb{R}^n$ are taken with respect to the n dimensional Hausdorff measure; those over its boundary with respect to the n-1 dimensional Hausdorff measure.

Radial capacity potentials can be expressed in terms of the *fundamental singularity of the p-Laplacian.* We denote it as

$$(2.4) \qquad K(r) \;:=\; \begin{cases} \frac{p-1}{n-p}\left|S^{n-1}\right|^{\frac{1}{1-p}} r^{\frac{p-n}{p-1}} & (p < n), \\ -\left|S^{n-1}\right|^{\frac{1}{1-n}} \log(r) & (p = n). \end{cases}$$

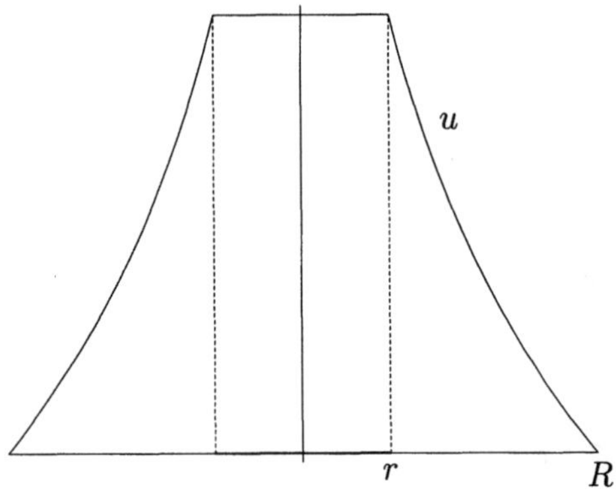


Figure 2.1: Radial capacity potential.

where $|S^{n-1}|$ denotes the surface area of the unit sphere in $\mathbb{R}^n$. The constants are obtained from the flux condition $|S^{n-1}|\,|K'(1)|^{p-1} = 1$ with

$$K'(r) \quad = \quad -|S^{n-1}|^{\frac{1}{1-p}}\, r^{\frac{1-n}{p-1}}. \tag{2.5}$$

In the semilinear case we have

$$K(r) \quad = \quad \begin{cases} -\frac{1}{2\pi}\log(r) & (n=2), \\ \frac{1}{(n-2)|S^{n-1}|r^{n-2}} & (n \geq 3). \end{cases}$$

The p-capacity potential of a ball $B_x^r$ of radius $r$ centered at $x$ with respect to the concentric ball $B_x^R$ is

$$u(y) \quad = \quad \frac{K(|y-x|) - K(R)}{K(r) - K(R)} \quad \text{for } \; r \leq |y-x| \leq R \tag{2.6}$$

extended by 1 in $B_0^r$ (Figure 2.1). The corresponding p-capacity is

$$\mathrm{pcap}_{B_0^R}(B_0^r) \quad = \quad (K(r) - K(R))^{1-p}. \tag{2.7}$$

In the conformal case

$$\mathrm{ncap}_{B_0^R}(B_0^r) \quad = \quad |S^{n-1}| \left(\log \frac{R}{r}\right)^{1-n}. \tag{2.8}$$

In contrast to the subconformal p-capacity the conformal n-capacity tends to 0 as $R \to \infty$. In the subconformal case the capacity satisfies the *isoperimetric inequality relating capacity and volume*

$$|A| \quad \leq \quad F^* \mathrm{pcap}_\Omega^{\frac{n}{n-p}}(A). \tag{2.9}$$

Equality holds if and only if $A$ is a ball and $\Omega = \mathbb{R}^n$. The optimal constant $F^*$ can be computed from this case (Section 3.3). The semilinear three dimensional inequality is due to Szegö [122]. Today the general inequality (2.9) is easily proved by means of Schwarz symmetrization (Kawohl [71]). The *Schwarz symmetrization* of a positive function $u \in D^{1,p}(\Omega)$ is denoted by $u^*$, the symmetrized domain by $\Omega^*$, its radius by $R_\Omega$. Note that constant boundary data is necessary for Schwarz symmetrization. For a negative function we define $u^* := -(-u)^*$. This transformation has the well known properties

$$\begin{aligned} \int_{\Omega^*} |\nabla u^*|^p &\leq \int_\Omega |\nabla u|^p, \\ \int_{\Omega^*} F(u^*) &= \int_\Omega F(u). \end{aligned}$$

In the conformal case (2.9) it assumes the form

$$|A| \leq |\Omega| \exp\left(-\alpha_n \mathrm{nmod}_\Omega(A)\right) \tag{2.10}$$

with

$$\alpha_n := n \left|S^{n-1}\right|^{\frac{1}{n-1}}. \tag{2.11}$$

The semilinear inequality goes back to Carleman [31]. Equality holds if and only if $\Omega$ and $A$ are concentric balls. The p-capacity of a connected set $A$ is bounded below by

$$\mathrm{pcap}_\Omega(A) \geq \mathrm{diam}(A)\, \mathrm{pcap}_{\mathbb{R}^{n-1}}(\{0\}) \tag{2.12}$$

as follows by symmetrization in $\mathbb{R}^{n-1}$. The right hand side is nontrivial for $p > n-1$. While the isoperimetric inequality gives a lower bound for the capacity the following lemma provides an upper bound. For simply connected planar domains with $p = 2$ it is known as *Grötzsch's lemma* (Teichmüller [124], Pólya and Schiffer [103]).

**Lemma 2.1 (Subadditivity of p-modulus)** *The relative p-moduli of nested domains $A \subset B \subset C$ satisfy the inequality*

$$\mathrm{pmod}_C(A) \geq \mathrm{pmod}_B(A) + \mathrm{pmod}_C(B).$$

*Equality holds if $B$ is a level set of the p-capacity potential of $A$ with respect to $C$.*

*Proof.* Any convex combination of the p-capacity potentials on the right is an admissible comparison function for the left side. Therefore

$$\mathrm{pcap}_C(A) \leq \theta^p \mathrm{pcap}_B(A) + (1-\theta)^p \mathrm{pcap}_C(B)$$

for every $\theta \in [0,1]$. Minimization over $\theta$ yields the assertion. □

Further properties of the p-capacity will be discussed in Section 3.3 on the volume integrand and Chapter 14 on Bernoulli's free-boundary problem.

# Chapter 3

# Generalized Sobolev Inequality

In this chapter we investigate the subconformal case $1 < p < n$. The conformal case is delayed to Chapter 12. The corresponding semilinear problems have been studied with S. Müller [55].

To initiate our analysis of low energy limits we make a simple but fundamental observation. By scaling of the independent variable the classical Sobolev inequality for the critical power integrand extends to arbitrary integrands satisfying the growth condition (1.3). Similarly we can define a generalized Sobolev constant.

**Definition 3.1 (Generalized Sobolev constant)**

$$F^* \quad := \quad \sup\left\{\int_{\mathbb{R}^n} F(w) \ : \ w \in D^{1,p}(\mathbb{R}^n), \ \|\nabla w\|_p \leq 1\right\}.$$

Our general hypotheses (1.3) and Sobolev's inequality (1.4) imply $0 < F^* < \infty$. Schwarz symmetrization reduces the computation of the generalized Sobolev constant to a one dimensional problem. The classical Sobolev inequality extends as follows.

**Lemma 3.2 (Generalized Sobolev inequality)** *If the growth condition* (1.3) *holds then*

$$\int_\Omega F(u) \quad \leq \quad F^* \|\nabla u\|_p^{p^*}$$

*for every domain $\Omega \subset \mathbb{R}^n$ and every $u \in D^{1,p}(\Omega)$.*

*Proof.* This is a simple consequence of the basic scaling properties of integrals in $\mathbb{R}^n$. Set

$$u^s(x) \quad := \quad u(x/s).$$

Then

$$\begin{aligned} \int_{s\Omega} F(u^s) &= s^n \int_\Omega F(u), \\ \int_{s\Omega} |\nabla u^s|^p &= s^{n-p} \int_\Omega |\nabla u|^p . \end{aligned} \tag{3.1}$$

We refer to this transformation as *scaling of the independent variable.* It can be used to normalize a function whose p-Dirichlet integral is not 1. If we choose $s = \|\nabla u\|_p^{-\frac{p}{n-p}}$ then $\|\nabla u^s\| = 1$ and by the definition of $p^*$ we obtain

$$\int_{s\Omega} F(u^s) = \|\nabla u\|_p^{-p^*} \int_\Omega F(u). \tag{3.2}$$

By definition of the generalized Sobolev constant the left hand side is at most $F^*$. □

The growth condition is crucial for finiteness of $F^*$. An integrand with

$$\frac{F(t)}{|t|^{p^*}} \longrightarrow \infty \quad \text{as} \quad t \to 0$$

does not satisfy a generalized Sobolev inequality as we will see in Proposition 7.2.

## 3.1 Local generalized Sobolev inequality

Using n-capacity potentials as cut-off functions we derive a local version of the generalized Sobolev inequality. This result will lead to simple and unified proofs of the generalized concentration compactness alternative (Theorem 4.11) and the concentration result (Theorem 7.3).

**Lemma 3.3 (N-harmonic cut-off functions)** *For every $\delta > 0$ there is a constant $k(\delta) > 0$ with the following property. If $0 < r < R$ with $r/R \le k(\delta)$ then there is a cut-off function $\eta_R^r \in D^{1,\infty}(\mathbb{R}^n)$ such that $\eta_R^r = 1$ in $B_x^r$, $\eta_R^r = 0$ outside $B_x^R$, and*

$$\int_{B_x^R} |\nabla(\eta_R^r u)|^p \le \int_{B_x^R} |\nabla u|^p + \delta \int_{\mathbb{R}^n} |\nabla u|^p$$

*for every $u \in D^{1,p}(\mathbb{R}^n)$.*

*Proof.* We may suppose $x = 0$. Let $\eta_R^r$ be the n-capacity potential of $B_0^r$ with respect to $B_0^R$. We use Young's inequality in the form

$$|a+b|^p \le (1+\beta)^{p-1} |a|^p + \left(1 + \frac{1}{\beta}\right)^{p-1} |b|^p \tag{3.3}$$

for every $a, b \in \mathbb{R}$ and arbitrary $\beta > 0$. Together with Hölder's and Sobolev's inequality we obtain

$$\begin{aligned}
\int_{B_0^R} |\nabla(\eta_R^r u)|^p &\leq (1+\beta)^{p-1} \int_{B_0^R} |\eta_R^r|^p \, |\nabla u|^p \\
&\quad + \left(1+\frac{1}{\beta}\right)^{p-1} \int_{B_0^R} |u|^p \, |\nabla \eta_R^r|^p \\
&\leq (1+\beta)^{p-1} \int_{B_0^R} |\nabla u|^p \\
&\quad + \left(1+\frac{1}{\beta}\right)^{p-1} \left(\int_{B_0^R} |u|^{p^*}\right)^{\frac{n-p}{n}} \left(\int_{B_0^R} |\nabla \eta_R^r|^n\right)^{p/n} \\
&\leq \int_{B_0^R} |\nabla u|^p + \delta(\beta) \int_{\mathbb{R}^n} |\nabla u|^p
\end{aligned}$$

with

$$\delta(\beta) = (1+\beta)^{p-1} - 1 + c^p \left(1+\frac{1}{\beta}\right)^{p-1} \left(\log \frac{R}{r}\right)^{-\frac{p(n-1)}{n}}$$

by the formula for the n-capacity of balls (2.8). With the optimal choice of $\beta$ we can achieve $\delta(\beta) \leq \delta$ provided that

$$\left(\log \frac{R}{r}\right)^{-\frac{p(n-1)}{n}} \leq \left(\frac{\sqrt{1+\delta}-1}{c}\right)^p.$$

Hence we can choose

$$\begin{aligned}
k(\delta) &:= \exp\left(-\left(\frac{c}{(1+\delta)^{1/p}-1}\right)^{\frac{n}{n-1}}\right) \\
&= \exp\left(-\left(\frac{cp}{\delta}\right)^{\frac{n}{n-1}}\right)(1+o(1))
\end{aligned} \tag{3.4}$$

as $\delta \to 0$. □

**Corollary 3.4 (Local generalized Sobolev inequality)** *Assume the growth condition* (1.3). *Fix $\delta > 0$ and $r/R \leq k(\delta)$ with $k(\delta)$ as in* (3.4). *Then*

$$\begin{aligned}
\int_{B_x^r} F(u) &\leq F^* \left(\int_{B_x^R} |\nabla u|^p + \delta \int_{\mathbb{R}^n} |\nabla u|^p\right)^{p^*/p}, \\
\int_{\mathbb{R}^n \setminus B_x^R} F(u) &\leq F^* \left(\int_{\mathbb{R}^n \setminus B_x^r} |\nabla u|^p + \delta \int_{\mathbb{R}^n} |\nabla u|^p\right)^{p^*/p}
\end{aligned}$$

*for every $u \in D^{1,p}(\mathbb{R}^n)$.*

*Proof.* We may suppose $x = 0$. Using the notation of Lemma 3.3 and the generalized Sobolev inequality we can estimate

$$\begin{aligned}\int_{B_0^r} F(u) &\le \int_{B_0^R} F(\eta_R^r u) \le F^* \|\nabla(\eta_R^r u)\|_p^{p^*} \\ &\le F^* \left( \int_{B_0^R} |\nabla u|^p + \delta \int_{\mathbb{R}^n} |\nabla u|^p \right)^{p^*/p} .\end{aligned}$$

For the second inequality we use the cut-off function $(1-\eta_R^r)\eta_{R_2}^{R_1}$ with $r < R < R_1 < R_2$ and let $R_1$ and $R_2 \to \infty$ such that $R_2/R_1 \to \infty$. □

The following examples will serve as model problems to illustrate our results.

## 3.2 Critical power integrand

For the critical power integrand

(3.5) $$S(t) := |t|^{p^*}$$

the generalized Sobolev inequality reduces to the classical Sobolev inequality (1.4). Usually

$$(S^*)^{\frac{p-n}{n}} = n\pi^{p/2} \left(\frac{n-p}{p-1}\right)^{p-1} \left(\frac{\Gamma(n/p)\Gamma(1+n-n/p)}{\Gamma(1+n/2)\Gamma(n)}\right)^{p/n}$$

is called the *best Sobolev constant.* Up to translation extremal functions for $S^*$ are of the form

(3.6) $$w(x) = \frac{c}{\left(s + |x|^{\frac{p}{p-1}} / s^{\frac{1}{p-1}}\right)^{\frac{n-p}{p}}}$$

with $s > 0$ (Talenti [123]). In the semilinear case in three dimensions the constant is $c = 2/(\sqrt{3}\pi)$.

## 3.3 Volume integrand

For the integrand

$$F(t) := \begin{cases} 0 & (t < 1), \\ 1 & (t \ge 1) \end{cases}$$

an extremal function $u$ of the variation problem (1.2) is the capacity potential of the set $A := \{u \ge 1\}$ with $\int_\Omega F(u) = |A|$ and $\int_\Omega |\nabla u|^p = \mathrm{pcap}_\Omega(A)$. Thus it suffices to solve the reduced problem

(3.7) $$\sup\{|A| : \mathrm{pcap}_\Omega(A) \le \varepsilon^p\} .$$

This form explains the term volume integrand. The weak form of the corresponding Euler Lagrange equation is Bernoulli's free-boundary problem (Chapter 14). The entire extremals for $F^*$ are the translates of

$$w(x) = \begin{cases} K(|x|) & (|x| > r), \\ 1 & (|x| \le r) \end{cases} \tag{3.8}$$

with $r$ such that $K(r) = 1$. Note that $\|\nabla w\|_p^p = \operatorname{pcap}_{\mathbb{R}^n}(B_0^r) = K(r)^{1-p} = 1$ by the formula for the capacity of balls (2.7). The corresponding generalized Sobolev constant is

$$F^* = \int_{\mathbb{R}^n} F(w) = |B_0^r|.$$

The semilinear three dimensional constant is $F^* = (48\pi^2)^{-1}$. The generalized Sobolev inequality for the volume integrand is the isoperimetric inequality relating p-capacity and volume (2.9).

## 3.4 Plasma integrand

In contrast to the volume integrand the *plasma integrand* is continuous. We consider

$$F(t) = \begin{cases} (t-1)^q & (t \ge 1), \\ 0 & (t < 1) \end{cases}$$

with *subcritical exponent* $q < 2^* = \frac{2n}{n-2}$. The corresponding Euler Lagrange equation is the free-boundary problem

$$\begin{aligned} -\Delta u &= \lambda(u-1)_+^{q-1} \text{ in } \Omega \setminus \partial A, \\ u &= 0 \text{ on } \partial\Omega, \\ u &= 1 \text{ on } \partial A, \\ u &\text{ is } C^1 \text{ across } \partial A. \end{aligned} \tag{3.9}$$

In plasma physics the domain $\Omega$ represents a two dimensional cross section of a tokamak, i.e. a toroidal shell containing a plasma ring surrounded by vacuum. The radius of the torus is large. We approximate it by a cylinder. The plasma consists of a perfectly conducting, ionized, ideal gas. The magnetic field $B$, the current $j$, and the pressure $p$ satisfy the equations of magnetohydrodynamics (MHD equations)

$$\operatorname{div} B = 0, \quad \operatorname{curl} B = j, \quad \nabla p = j \times B.$$

We consider the component of the magnetic field which is parallel to the cross section $\Omega$. By the first equation it has a scalar potential $u$, i.e. $B = \left(-\frac{\partial u}{\partial x_2}, \frac{\partial u}{\partial x_1}\right)$.

In the vacuous part $\Omega \setminus A$ the magnetic field is curl free and the potential is harmonic. In the plasma region $A$ the induced current $j$ leads to a motion of the plasma (ions). The magnetic field is tangential to the interface between plasma and vacuum. By the third equation the pressure gradient is orthogonal to the interface. An additional strong magnetic field is needed to compensate for this force and to confine the plasma. The simplest physical model including a constitutive law for the plasma leads to a problem of the form (3.9) for the magnetic potential (Temam [125]). The entire extremals for the plasma problem in the three dimensional linear case $p = q = 2$ (Figure 1.2 on page 3) are the translates of

$$w(x) = \begin{cases} \frac{r}{|x|} & (|x| > r), \\ 1 + \frac{r}{\pi|x|} \sin\left(\frac{\pi|x|}{r}\right) & (|x| \leq r) \end{cases} \tag{3.10}$$

with $r = (6\pi)^{-1}$. This leads to $F^* = \int_{\mathbb{R}^n} F(w) = (108\pi^4)^{-1}$.

# Chapter 4

# Concentration Compactness Alternatives

Most of the technical material of this monograph is contained in the chapters 4 trough 12. The reader who is mainly interested in the applications may skip these at first reading.

The concentration compactness alternatives of this chapter will be used in two ways. In a first set of applications we prove existence of extremal functions by exclusion of concentration (Chapter 5). In the second set of applications we prove concentration of low energy extremals by excluding compactness (Section 7.1). The corresponding semilinear results have been obtained with S. Müller [55].

In the following the closure $\overline{\Omega}$ of a domain is always taken in $\mathbb{R}^n \cup \{\infty\}$. In particular the closure of an unbounded domain contains the point $\infty$ and we allow for concentration at infinity. $\mathcal{M}(\overline{\Omega})$ denotes the class of non-negative Borel measures of finite total mass. *Convergence in the sense of measures*

$$\mu_\varepsilon \stackrel{*}{\rightharpoonup} \mu \text{ in } \mathcal{M}(\overline{\Omega})$$

is defined by $\int_{\overline{\Omega}} \eta \, d\mu_\varepsilon \to \int_{\overline{\Omega}} \eta \, d\mu$ for every test function $\eta \in C^\infty(\Omega) \cap C(\overline{\Omega})$. To detect the asymptotic behaviour of the extremals $u_\varepsilon$ we analyze the limit of the normalized measures

$$\mu_\varepsilon \;:=\; \frac{|\nabla u_\varepsilon|^p}{\int_\Omega |\nabla u_\varepsilon|^p} \;\stackrel{*}{\rightharpoonup}\; \mu \text{ in } \mathcal{M}(\overline{\Omega})$$

which exists for a subsequence. A point $x$ is called an *atom* if $\mu(\{x\}) > 0$. A Dirac measure $\mu = \delta_x$ has a single atom of unit mass. If $\mu_\varepsilon \stackrel{*}{\rightharpoonup} \delta_x$ we say that the sequence $(u_\varepsilon)$ concentrates at the point $x$.

**Example 4.1** [Concentration of radial capacity potentials] As an example of single point concentration we consider the radial p-capacity potentials $u_\varepsilon$ of

$A_\varepsilon := B_0^{r_\varepsilon}$ with respect to $\Omega := B_0^R$ given by (2.6) with $\varepsilon^p = (K(r_\varepsilon) - K(R))^{1-p}$. In $\Omega \setminus A_\varepsilon$ the normalized functions $v_\varepsilon := u_\varepsilon/\varepsilon$ satisfy

$$|\nabla v_\varepsilon(x)|^p = \frac{|K'(|x|)|^p}{K(r_\varepsilon) - K(R)}.$$

Away from the origin these functions tends uniformly to zero as $\varepsilon \to 0$. Thus $|\nabla v_\varepsilon(x)|^p \overset{*}{\rightharpoonup} \delta_0$ as $\varepsilon \to 0$. A subsequence converges weakly $v_\varepsilon \rightharpoonup v$ weakly in $D^{1,p}(\Omega)$. We show that the weak limit is $v = 0$. Using integration by parts and the boundary integral representation (2.3) we obtain

$$\begin{aligned} \int_\Omega |\nabla v_\varepsilon|^{p-2} \nabla v_\varepsilon \cdot \nabla \eta &= \int_{\partial A_\varepsilon} |\nabla v_\varepsilon|^{p-1} \eta \\ &\leq \frac{\|\eta\|_\infty}{\varepsilon^{p-1}} \mathrm{pcap}_\Omega(A_\varepsilon) = \varepsilon \|\eta\|_\infty \end{aligned}$$

for every test function $\eta \in C_c^\infty(\Omega)$. Together with the same estimate with $\eta$ replaced by $-\eta$ we conclude $v = 0$. This argument is not restricted to the radial case.

## 4.1 Concentration compactness alternative for critical power integrand

For the readers convenience we start by recalling the classical concentration compactness alternative of Lions [91, p. 158] for the critical power integrand in a form adapted to our purpose. This allows us to explain the differences and the similarities with our alternatives for general integrands. It is well known that the critical Sobolev embedding is noncompact on bounded domains. Instead every maximizing sequence concentrates at one point as follows from the alternative of Lions. Each alternative is split into two parts. A preliminary lemma describes the form of the limit measures achievable by arbitrary sequences. It contains the hard part of the proof. The actual alternative applies to maximizing sequences. It is stated in a separate theorem.

**Lemma 4.2** *Consider a sequence $(w_\varepsilon)$ in $D^{1,p}(\Omega)$ with*

$$\begin{aligned} \|\nabla w_\varepsilon\|_p &\leq 1, \\ w_\varepsilon &\rightharpoonup w \quad \text{weakly in } D^{1,p}(\Omega), \\ |\nabla w_\varepsilon|^p &\overset{*}{\rightharpoonup} \mu \quad \text{in } \mathcal{M}(\overline{\Omega}), \\ |w_\varepsilon|^{p^*} &\overset{*}{\rightharpoonup} \nu^* \quad \text{in } \mathcal{M}(\overline{\Omega}). \end{aligned}$$

*Then the limit measures are of the form*

$$\mu = |\nabla w|^p + \sum_{j=1}^J \mu_j \delta_{x_j} + \tilde{\mu}, \quad \mu(\overline{\Omega}) \leq 1,$$

$$\nu^* = |w|^{p^*} + \sum_{j=1}^J \nu_j^* \delta_{x_j}, \quad \nu^*(\overline{\Omega}) \leq S^* \mu(\overline{\Omega})^{p^*/p}$$

*with $J \in \mathbb{N} \cup \{\infty\}$ and a nonatomic positive measure $\widetilde{\mu} \in \mathcal{M}(\overline{\Omega})$. Moreover*

$$|w_\varepsilon - w|^{p^*} \stackrel{*}{\rightharpoonup} \sum_{j=1}^{J} \nu_j^* \delta_{x_j} \quad in \quad \mathcal{M}(\overline{\Omega}).$$

*The atoms satisfy the local Sobolev inequality*

$$0 \leq \nu_j^* \leq S^{\mathrm{loc}}(x_j)\mu_j^{p^*/p}$$

*with $S^{\mathrm{loc}}(x) = S^*$.*

We will consider more general alternatives in which the local Sobolev constant $S^{\mathrm{loc}}(x)$ is different from the global constant $S$. Clearly we always have $S^{\mathrm{loc}}(x) \leq S$.

**Theorem 4.3 (CCA for critical power integrand)** *Assume the hypotheses of Lemma 4.2. If in addition $(w_\varepsilon)$ is a maximizing sequence for $S^*$, i.e. $\nu^*(\overline{\Omega}) = S^*$ then $\mu(\overline{\Omega}) = 1$ and at least one of the following statements is true.*

*A Concentration: $\mu$ and $\nu^*$ are concentrated at a single point, i.e. $w = 0$, $\mu = \delta_{x_0}$, and $\nu = S^{\mathrm{loc}}(x_0)\delta_{x_0}$ for some $x_0 \in \overline{\Omega}$.*

*B Compactness: $w_\varepsilon \to w$ in $D^{1,p}(\Omega)$, $\mu = |\nabla w|^p$, and $\nu^* = |w|^{p^*}$.*

The key point in every subconformal concentration compactness alternative is the following convexity argument. With $\mu_0 := \int_\Omega |\nabla w|^p$ and $\nu_0 := \int_\Omega |w|^{p^*}$ one has

$$(4.1) \qquad \left(\sum_{j=0}^{J} \mu_j\right)^{p^*/p} \leq \mu(\overline{\Omega})^{p^*/p} = \frac{1}{S}\sum_{j=0}^{J} \nu_j^*$$
$$\leq \mu_0^{p^*/p} + \sum_{j=1}^{J} \frac{S^{\mathrm{loc}}(x_j)}{S} \mu_j^{p^*/p} \leq \sum_{j=0}^{J} \mu_j^{p^*/p}.$$

Note that for maximizing sequences $\nu^*(\overline{\Omega}) \leq S^*\mu(\overline{\Omega})^{p^*/p}$ is an equality. Moreover the nonatomic part of $\nu^*$ is regular. By strict convexity of the function $t \mapsto t^{p^*/p}$ on $\mathbb{R}^+$ all but one of the $\mu_j$ vanish. In fact $\mu_j = \delta_{kj}$ for some $k$. If $\mu_0 = 1$ we have compactness, otherwise concentration at a single point. Moreover (4.1) shows that concentration at a point $x$ can only occur if $S^{\mathrm{loc}}(x) = S$. This observation is the basis of the compactness criteria of Chapter 5.

The concentration compactness alternative for the critical power applies to maximizing sequences for the best Sobolev constant $S^*$. Assume $\Omega \neq \mathbb{R}^n$ up to a set of positive p-capacity, i.e. $\mathrm{pcap}_{\mathbb{R}^n}(\mathbb{R}^n \setminus \Omega) > 0$. Let $(w_\varepsilon)$ be a maximizing sequence for $S^*$. Then $\nu^*(\overline{\Omega}) = S^*$. If alternative B would hold then the best Sobolev constant would be attained by a function vanishing on $\mathbb{R}^n \setminus \Omega$. This is impossible because every extremal for $S^*$ is supported in all of $\mathbb{R}^n$. In fact it is of the form (3.6). We concluded that alternative A must hold, i.e. that a subsequence of $(w_\varepsilon)$ concentrates at a single point $x_0 \in \overline{\Omega}$. In addition $\mu = \delta_{x_0}$ and $\nu^* = S^*\delta_{x_0}$. On the other hand it is possible to construct a maximizing sequence for $S^*$ concentrating at an arbitrary prescribed point $x_0 \in \overline{\Omega}$.

## 4.2 Generalized concentration compactness alternative

In the generalized concentration compactness alternative the critical power integrand is replaced by a general integrand satisfying the growth condition (1.3). We will use it to prove convergence of low energy extremals on a microscopic scale near the concentration point (Section 7.2). The first two steps in the proof of the lemma are identical with the proof for the critical power integrand. The remaining steps are different.

**Lemma 4.4** *Assume the growth condition* (1.3). *Let* $(w_\varepsilon)$ *be a sequence in* $D^{1,p}(\Omega)$ *of norm* $\|\nabla w_\varepsilon\|_p \leq 1$ *such that*

$$\begin{aligned} w_\varepsilon &\rightharpoonup w \text{ weakly in } D^{1,p}(\Omega), \\ |\nabla w_\varepsilon|^p &\overset{*}{\rightharpoonup} \mu \text{ in } \mathcal{M}(\overline{\Omega}), \\ F(w_\varepsilon) &\overset{*}{\rightharpoonup} \nu \text{ in } \mathcal{M}(\overline{\Omega}). \end{aligned}$$

*Then the limit measures are of the form*

$$\begin{aligned} \mu &= |\nabla w|^p + \sum_{j=1}^{J} \mu_j \delta_{x_j} + \widetilde{\mu}, \quad \mu(\overline{\Omega}) \leq 1 \\ \nu &= g + \sum_{j=1}^{J} \nu_j \delta_{x_j}, \quad \nu(\overline{\Omega}) \leq F^* \end{aligned}$$

*with* $J \in \mathbb{N} \cup \{\infty\}$, *a nonatomic positive measure* $\widetilde{\mu} \in \mathcal{M}(\overline{\Omega})$, *and* $g \in L^1(\Omega)$. *The atoms and the regular part satisfy the generalized Sobolev inequality*

$$\begin{aligned} \nu(\overline{\Omega}) &\leq F^* \mu(\overline{\Omega})^{p^*/p}, \\ \nu_j &\leq F^* \mu_j^{p^*/p}, \\ \int_\Omega g &\leq F^* \left( \int_\Omega |\nabla w|^p + \widetilde{\mu}(\overline{\Omega}) \right)^{p^*/p}. \end{aligned}$$

*Proof.* For a subsequence we have

$$|w_\varepsilon|^{p^*} \overset{*}{\rightharpoonup} \nu^* \text{ in } \mathcal{M}(\overline{\Omega}).$$

We prove the statements about the form of each measure.

**Step 4.5 (Decomposition of $\mu$)** *The limit measure* $\mu$ *is of the form*

$$\mu = |\nabla w|^p + \sum_{j=1}^{J} \mu_j \delta_{x_j} + \widetilde{\mu}.$$

*with* $\mu(\overline{\Omega}) \leq 1$, $J \in \mathbb{N} \cup \{\infty\}$, *and a nonatomic positive measure* $\widetilde{\mu} \in \mathcal{M}(\overline{\Omega})$.

*Proof.* From every measure we can extract its atoms. Therefore it suffices to show that $\mu \geq |\nabla w|^p$. For every test function $\eta$ with compact support in $\overline{\Omega}$ we have

$$\int_\Omega |\nabla(\eta w)|^p \quad \leq \quad \liminf_{\varepsilon\to 0} \int_\Omega |\nabla(\eta w_\varepsilon)|^p$$

by weak lower semi continuity of the norm. Rellich's compactness theorem implies

$$\int_\Omega |\nabla w|^p \, |\eta|^p \quad \leq \quad \liminf_{\varepsilon\to 0} \int_\Omega |\nabla w_\varepsilon|^p \, |\eta|^p .$$

Hence $\mu \geq |\nabla w|^p$ and $\widetilde{\mu} \geq 0$. The set of atoms is countable because $\mu$ has finite total mass. □

**Step 4.6 (Decomposition of $\nu^*$)** *The measure $\nu^*$ is of the form*

$$\nu^* \quad = \quad |w|^{p^*} + \sum_{j=1}^{J} \nu_j^* \delta_{x_j}$$

*with $\nu(\overline{\Omega}) \leq S^*$. Moreover*

$$|w_\varepsilon - w|^{p^*} \quad \stackrel{*}{\rightharpoonup} \quad \sum_{j=1}^{J} \nu_j^* \delta_{x_j} \quad in \quad \mathcal{M}(\overline{\Omega}).$$

*Proof.* For a subsequence we have $w_\varepsilon \to w$ a.e. By the Brézis-Lieb lemma [26]

$$\liminf_{\varepsilon\to 0} \int_{\Omega'} |w_\varepsilon|^{p^*} - \int_{\Omega'} |w|^{p^*} \quad = \quad \liminf_{\varepsilon\to 0} \int_{\Omega'} |w_\varepsilon - w|^{p^*}$$

for every bounded subdomain $\Omega' \subset \Omega$. Thus

$$\overline{\nu} \quad := \quad \nu^* - |w|^{p^*} \quad = \quad \sum_{j=1}^{J} \nu_j^* \delta_{x_j} + \widetilde{\nu}$$

with nonatomic $\widetilde{\nu} \in \mathcal{M}(\overline{\Omega})$. By the generalized Sobolev inequality for the atoms (Step 4.9 below) applied to the critical power integrand every atom of $\nu^*$ is one of $\mu$. We show that $\widetilde{\nu} = 0$. Consider $\overline{w}_\varepsilon := w_\varepsilon - w$. For a subsequence we have $\overline{w}_\varepsilon \rightharpoonup 0$ weakly in $D^{1,p}(\Omega)$ and $|\nabla \overline{w}_\varepsilon|^p \stackrel{*}{\rightharpoonup} \overline{\mu}$ in $\mathcal{M}(\overline{\Omega})$. Let $\eta$ be the p-capacity potential of $B_x^r$ with respect to $B_x^R$. By Sobolev's inequality

$$\int_{B_x^R} |\eta \overline{w}_\varepsilon|^{p^*} \quad \leq \quad S^* \left( \int_{B_x^R} |\nabla(\eta \overline{w}_\varepsilon)|^p \right)^{p^*/p} .$$

As $\varepsilon \to 0$ we obtain

$$\overline{\nu}(\overline{B_x^r}) \quad \leq \quad \int_{\overline{B_x^R}} |\eta|^{p^*} \, d\overline{\nu} \quad \leq \quad S^* \left( \int_{\overline{B_x^R}} |\eta|^p \, d\overline{\mu} \right)^{p^*/p}$$

by (3.3), Rellich's compactness theorem, and $\overline{w}_\varepsilon \to 0$ in $L^p_{\mathrm{loc}}(\Omega)$. Letting $R \to r$ we obtain

$$\overline{\nu}(B_x^R) \;\leq\; S^* \left(\overline{\mu}(B_x^R)\right)^{p^*/p}. \tag{4.2}$$

In particular $\overline{\nu}$ is absolutely continuous with respect to $\overline{\mu}$. By the Radon-Nikodym theorem $\overline{\nu} = h\overline{\mu}$ with $h \in L^1(\overline{\Omega}, \overline{\mu})$. By Lebesgue's differentiation theorem

$$h(x) \;=\; \lim_{r\to 0} \frac{\overline{\nu}(\overline{B_x^r})}{\overline{\mu}(\overline{B_x^r})} \;\leq\; S^* \lim_{r\to 0} \overline{\mu}(\overline{B_x^r})^{\frac{p}{n-p}} \;=\; S^*\overline{\mu}(\{x\})^{\frac{p}{n-p}}$$

by (4.2). The right hand side vanishes except if $x$ is an atom of $\overline{\mu}$. In the latter case $h(x) = 0$ except if $x$ is also an atom of $\overline{\nu}$. By the Brézis-Lieb lemma the atoms of $\overline{\nu}$ and $\nu^*$ are the same, namely $\{x_1, \ldots, x_J\}$. We conclude that $h = 0$ as a function in $L^1(\overline{\Omega} \setminus \{x_1, \ldots, x_J\}, d\overline{\mu})$. Therefore $\overline{\nu} = 0$ in $\Omega \setminus \{x_1, \ldots, x_J\}$. Thus $\widetilde{\nu} = 0$ as desired. □

**Step 4.7 (Decomposition of $\nu$)** *The measure $\nu$ is of the form*

$$\nu \;=\; g + \sum_{j=1}^{J} \nu_j \delta_{x_j}$$

*with $g \in L^1(\Omega)$.*

*Proof.* By the growth condition (1.3) we have $0 \leq \nu \leq c\nu^*$. Application of the Radon Nikodym theorem with respect to $\nu^*$ yields $g \in L^1(\Omega)$. □

**Step 4.8 (Generalized Sobolev inequality for total mass)** *The integrals of $\mu$ and $\nu$ satisfy*

$$\nu(\overline{\Omega}) \;\leq\; F^* \mu(\overline{\Omega})^{p^*/p}.$$

*Proof.* By the generalized Sobolev inequality

$$\int_\Omega F(w_\varepsilon) \;\leq\; F^* \left(\int_\Omega |\nabla w_\varepsilon|^p\right)^{p^*/p}.$$

Passing to the limit yields the assertion since $\eta = 1$ is an admissible test function for convergence in $\mathcal{M}(\overline{\Omega})$. □

**Step 4.9 (Generalized Sobolev inequality for atoms)** *For every $x \in \overline{\Omega}$ one has*

$$\nu(\{x\}) \;\leq\; F^* \mu(\{x\})^{p^*/p}.$$

*In particular $\nu_j \leq F^* \mu_j^{p^*/p}$ for every $j \geq 1$.*

*Proof.* Let $\delta > 0$ and $R > 0$. If $x \in \mathbb{R}^n$ it follows from the local generalized Sobolev inequality (Corollary 3.4) that

$$\nu(\{x\}) \;\leq\; \nu\left(\overline{B_x^r}\right) \;\leq\; F^* \left(\mu\left(\overline{B_x^R}\right) + \delta\right)^{p^*/p}$$

for $r \leq k(\delta)R$. The assertion follows as $r \to 0$, $\delta \to 0$, and $R \to 0$. If $x = \infty$ we apply the second inequality of Corollary 3.4. □

**Step 4.10 (Generalized Sobolev inequality for regular part)** *The regular part $g$ of $\nu$ satisfies*

$$\int_\Omega g \;\le\; F^*\left(\int_\Omega |\nabla w|^p + \widetilde{\mu}(\overline{\Omega})\right)^{\frac{n}{n-p}}.$$

*Proof.* To cut out the atoms fix $\delta > 0$, $R_1 > 0$, and a finite $I \le J$ such that $\sum_{j=I+1}^J \mu_j \le \delta$. Choose $R > 0$ so small that the balls $B^R_{x_1}, \ldots, B^R_{x_I}$ are disjoint. The cut-off function

$$\eta \;:=\; \eta^{R_1}_{R_2} \prod_{i=1}^I (1 - \eta^r_R(\cdot - x_i))$$

is supported in $B^{R_2}_0 \setminus \bigcup_{i=1}^I B^r_{x_i}$. The regular part of $\nu$ satisfies $g \le \nu$. By the local generalized Sobolev inequality there exist $r > 0$ and $R_2 > R_1$ such that

$$\begin{aligned}
\int_{B^{R_1}_0 \setminus \bigcup_{i=1}^I B^R_{x_i}} g \;&\le\; \nu\left(\overline{B^{R_1}_0 \setminus \bigcup_{i=1}^I B^R_{x_i}}\right) \\
&\le\; F^*\left(\mu\left(\overline{B^{R_2}_0 \setminus \bigcup_{i=1}^I B^r_{x_i}}\right) + \delta\right)^{p^*/p} \\
&\le\; F^*\left(\int_\Omega |\nabla w|^p + \widetilde{\mu}(\overline{\Omega}) + \delta\right)^{\frac{n}{n-p}}.
\end{aligned}$$

Then let $R \to 0$, $R_1 \to \infty$, and $\delta \to 0$. □

□

**Theorem 4.11 (Generalized CCA)** *Assume the hypotheses of Lemma 4.4. If in addition $\nu(\overline{\Omega}) = F^*$ then $\mu(\overline{\Omega}) = 1$ and at least one of the following statements is true.*

- *A Concentration: $w = 0$, $\mu = \delta_{x_0}$, and $\nu = F^*\delta_{x_0}$ for some $x_0 \in \overline{\Omega}$.*
- *B Compactness: $w_\varepsilon \to w$ in $D^{1,p}(\Omega)$, $w$ is an extremal for $F^*$, $\mu = |\nabla w|^p$, and $F(w_\varepsilon) \to F(w)$ in $L^1(\Omega)$.*

*Compactness can only occur if $\Omega = \mathbb{R}^n$ up to a set of $p$-capacity zero.*

*Proof.* Set $\nu_0 := \int_\Omega g$ and $\mu_0 := \int_\Omega |\nabla w|^p + \widetilde{\mu}(\overline{\Omega})$. By Step 4.8 and normalization of $w_\varepsilon$ we deduce

$$F^* \;=\; \nu(\overline{\Omega}) \;\le\; F^*\mu(\overline{\Omega})^{p^*/p} \;\le\; F^*.$$

Hence $\mu(\overline{\Omega}) = 1$. By the Steps 4.8, 4.9, and 4.10 the convexity argument (4.1) applies with $S = S^{\text{loc}}(x) = F^*$. If $\mu_0 = 0$ then alternative A holds. If $\mu_0 \neq 0$, then $\nu_j^* = \nu_j = \mu_j = 0$ for every $j \geq 1$, $\mu_0 = 1$, and $\int_\Omega g = \nu_0 = F^*$. For a subsequence we have $w_\varepsilon \to w$ a.e. We show that alternative B holds in this case. By upper semi continuity of $F$ and application of Fatou's lemma to the sequence $(c\,|w_\varepsilon|^{p^*} - F(w_\varepsilon))$ it follows that

$$\begin{aligned} F^* &= \lim_{\varepsilon\to 0}\int_\Omega F(w_\varepsilon) \leq \int_\Omega \limsup_{\varepsilon\to 0} F(w_\varepsilon) \\ &\leq \int_\Omega F(w) \leq F^*\|\nabla w\|_p^{p^*} \leq F^*. \end{aligned}$$

Equality in the above chain of inequalities implies $\|\nabla w\|_p = 1$ and $\int_\Omega F(w) = F^*$. Together with weak convergence this implies strong convergence of $w_\varepsilon \to w$ in $D^{1,p}(\Omega)$. Moreover

$$\limsup_{\varepsilon\to 0} F(w_\varepsilon) = F(w) \text{ a.e.} \tag{4.3}$$

By Step 4.6 we know that $\nu^* = |w|^{p^*}$ and $\int_{\Omega'} |w_\varepsilon|^{p^*} \to \int_{\Omega'} |w|^{p^*}$ for every bounded subdomain $\Omega'$. Together with weak convergence we get $w_\varepsilon \to w$ in $L^{p^*}_{\text{loc}}(\Omega)$. The measure $\mu$ is nonatomic. In particular $\mu(\{\infty\}) = 0$. Hence $\nu^*(\{\infty\}) = 0$ and $w_\varepsilon \to w$ in $L^{p^*}(\Omega)$. By the growth condition and $|w_\varepsilon|^{p^*} \to |w|^{p^*}$ in $L^1(\Omega)$ the functions $(F(w_\varepsilon))$ are equi-integrable. By the Dunford-Pettis compactness criterion a bounded sequence of equi-integrable $L^1$-functions is weakly compact (see e.g. Dellacherie and Meyer [38, Theorem 25]). In our case

$$F(w_\varepsilon) \rightharpoonup g \text{ weakly in } L^1(\Omega) \tag{4.4}$$

for a subsequence. Moreover $g \leq F(w)$ a.e. by upper semi continuity of $F$. On the other hand $\int_\Omega F(w) = F^* = \int_\Omega g$ and thus $g = F(w)$ a.e. By (4.3) we have $(F(w_\varepsilon) - F(w))^+ \to 0$ a.e. Since $0 \leq F(w_\varepsilon) \leq c\,|w_\varepsilon|^{p^*} \to c\,|w|^{p^*}$ in $L^1(\Omega)$ Lebesgue's dominated convergence theorem implies $(F(w_\varepsilon) - F(w))^+ \to 0$ in $L^1(\Omega)$. By (4.4) the same is true for the negative part:

$$\int_\Omega (F(w_\varepsilon) - F(w))^- = \int_\Omega (F(w_\varepsilon) - F(w)) - \int_\Omega (F(w_\varepsilon) - F(w))^+ \to 0.$$

Hence $F(w_\varepsilon) \to F(w)$ in $L^1(\Omega)$ as desired. Every extremal for $F^*$ is either strictly positive or strictly negative in all of $\mathbb{R}^n$ (Lemma 6.2). Thus compactness can only occur if $\Omega = \mathbb{R}^n$ up to a set of p-capacity zero. □

## 4.3 Concentration compactness alternative for low energy extremals

Concentration of low energy extremals (Theorem 7.3) will be proved by means of the alternative of this section. Formally the alternative of this section looks similar to that of the previous section. However there are some subtle differences. The following quantities will be used to compare $F$ to the critical power integrand.

$$\begin{aligned} F_0^+ &:= \limsup_{t\to 0} \frac{F(t)}{|t|^{p^*}}, & F_0^- &:= \liminf_{t\to 0} \frac{F(t)}{|t|^{p^*}}, \\ F_\infty^+ &:= \limsup_{|t|\to \infty} \frac{F(t)}{|t|^{p^*}}, & F_\infty^- &:= \liminf_{|t|\to \infty} \frac{F(t)}{|t|^{p^*}}. \end{aligned}$$

In the case of equality we write $F_0 := F_0^+ = F_0^-$ and $F_\infty := F_\infty^+ = F_\infty^-$.

**Lemma 4.12** *Assume the growth condition* (1.3). *Let* $(u_\varepsilon)$ *be a sequence in* $D^{1,p}(\Omega)$ *with* $\|\nabla u_\varepsilon\|_p \leq \varepsilon$ *and define* $v_\varepsilon := u_\varepsilon/\varepsilon$. *Assume*

$$\begin{aligned} v_\varepsilon &\rightharpoonup v \text{ weakly in } D^{1,p}(\Omega), \\ |\nabla v_\varepsilon|^p &\overset{*}{\rightharpoonup} \mu \text{ in } \mathcal{M}(\overline{\Omega}), \\ \varepsilon^{-p^*} F(u_\varepsilon) &\overset{*}{\rightharpoonup} \nu \text{ in } \mathcal{M}(\overline{\Omega}). \end{aligned}$$

*Then the limit measures are of the form*

$$\begin{aligned} \mu &= |\nabla v|^p + \sum_{j=1}^J \mu_j \delta_{x_j} + \widetilde{\mu}, \quad \mu(\overline{\Omega}) \leq 1 \\ \nu &= g + \sum_{j=1}^J \nu_j \delta_{x_j}, \quad \nu(\overline{\Omega}) \leq F^* \mu(\overline{\Omega})^{p^*/p} \end{aligned}$$

*with* $J \in \mathbb{N} \cup \{\infty\}$, *a nonatomic positive measure* $\widetilde{\mu} \in \mathcal{M}(\overline{\Omega})$, *and* $g \in L^1(\Omega)$ *such that*

$$\begin{aligned} \nu_j &\leq F^* \mu_j^{p^*/p}, \\ \int_\Omega g &\leq F^* \left( \int_\Omega |\nabla v|^p + \widetilde{\mu}(\overline{\Omega}) \right)^{p^*/p}, \\ g &\leq F_0^+ |v|^{p^*} \quad \text{a.e. in } \Omega, \\ \int_\Omega g &\leq F_0^+ S^* \left( \int_\Omega |\nabla v|^p \right)^{p^*/p}. \end{aligned}$$

*Proof.* The proof of Lemma 4.12 is similar to that of Lemma 4.4 with the following exception. We need a pointwise estimate for the regular part of $\nu$. We can prove such an estimate using the fact that the large values of the $v_\varepsilon$ do not contribute to the regular part. Let $\Omega'$ be a subdomain. For $t > 0$ we have

$$\begin{aligned}\int_{\Omega'} g &\le \nu(\Omega') \le \varepsilon^{-p^*} \liminf_{\varepsilon\to 0} \int_{\Omega'} F(u_\varepsilon) \\ &\le \varepsilon^{-p^*} \limsup_{\varepsilon\to 0} \int_{\Omega'\cap\{|v_\varepsilon|<t\}} F(u_\varepsilon) + c \limsup_{\varepsilon\to 0} \int_{\Omega'\cap\{|v_\varepsilon|\ge t\}} |v_\varepsilon|^{p^*} \\ &\le (F_0^+ + o(1))\nu^*(\overline{\Omega'}) + c \limsup_{\varepsilon\to 0} \left( \int_{\Omega'\cap\{|v_\varepsilon|\ge t\}} |v|^{p^*} + \int_{\Omega'} |v_\varepsilon - v|^{p^*} \right).\end{aligned}$$

By the isoperimetric inequality for the p-capacity (2.9) and normalization of $v_\varepsilon$ we know that

$$|\{|v_\varepsilon| \ge t\}| \le c\, \mathrm{pcap}_\Omega^{\frac{n}{n-p}}(\{|v_\varepsilon| \ge t\}) \le c t^{-p^*} \to 0$$

as $t \to \infty$. In combination with the second assertion of Step 4.6 we conclude that

$$\int_{\Omega'} g \le F_0^+ \nu^*(\overline{\Omega'}) + c \sum_{x_j\in\overline{\Omega'}} \nu_j^* \le F_0^+ \int_{\Omega'} |v|^{p^*} + (1+c) \sum_{x_j\in\overline{\Omega'}} \nu_j^*.$$

Application of the Radon-Nikodym theorem yields $g \le F_0^+ |v|^{p^*}$ a.e. The remaining inequalities of the lemma follow by integration, Sobolev's inequality, and $F_0^- S^* \le F^*$. □

**Theorem 4.13 (CCA for low energy extremals)** *Assume the hypotheses of Lemma 4.12. If in addition $\nu(\overline{\Omega}) = F^*$ then $\mu(\overline{\Omega}) = 1$ and at least one of the following statements is true.*

*A Concentration: $v = 0$, $\mu = \delta_{x_0}$, and $\nu = F^*\delta_{x_0}$ for some $x_0 \in \overline{\Omega}$.*

*B Compactness:*

$$\begin{aligned}\mu &= |\nabla v|^p + \tilde{\mu}, \\ v_\varepsilon &\to v \text{ in } L^{p^*}(\Omega), \quad F^* \le c \int_\Omega |v|^{p^*}, \\ \varepsilon^{-p^*} F(u_\varepsilon) &\rightharpoonup g \text{ weakly in } L^1(\Omega), \quad F^* = \int_\Omega g \le F_0^+ S^*.\end{aligned}$$

*If in addition $F_0^+ = F_0^-$ then $v_\varepsilon \to v$ in $D^{1,p}(\Omega)$, $F^* = F_0 S^*$, $v$ is an extremal for $S^*$ and $\Omega = \mathbb{R}^n$ up to a set of p-capacity zero.*

*Proof.* The proof of Theorem 4.13 is similar to that of the generalized concentration compactness alternative (Theorem 4.11) with the following exceptions. Replacing $F(w_\varepsilon)$ by $\varepsilon^{-p^*} F(u_\varepsilon)$ in the proof of Theorem 4.11 we obtain that either concentration occurs or $\varepsilon^{-p^*} F(u_\varepsilon) \rightharpoonup g$ weakly in $L^1(\Omega)$. However, this time we do not get strong $L^1$-convergence of $\varepsilon^{-p^*} F(u_\varepsilon)$ although we know that

$\int_\Omega g = F^* = \lim \varepsilon^{-p^*} \int_\Omega F(u_\varepsilon)$. The reason is that the $L^1$-norm is not strictly convex.

The assertion for the case $F_0^+ = F_0^-$ is obtained as follows. We may assume $F_0 > 0$ since otherwise $F^* = 0$. By Lemma 4.12

$$F^* \;=\; \nu_0 \;\leq\; F^* \left( \int_\Omega |\nabla v|^p \right)^{p^*/p}.$$

Hence $\|\nabla v\|_p = 1$. Since $\nabla v_\varepsilon \rightharpoonup \nabla v$ weakly in $L^p(\Omega)$ it follows that $v_\varepsilon \to v$ strongly in $D^{1,p}(\Omega)$. We show that $(v_\varepsilon)$ is a maximizing sequence for $S^*$. Fix $t > 0$. We estimate

$$\begin{aligned} \varepsilon^{-p^*} \int_\Omega F(u_\varepsilon) &\leq \varepsilon^{-p^*} \int_{\{|v_\varepsilon| \leq t\}} F(\varepsilon v_\varepsilon) + c \int_{\{|v_\varepsilon| \geq t\}} |v_\varepsilon|^{p^*} \\ &\leq (F_0^+ + o(1)) \int_\Omega |v_\varepsilon|^{p^*} + c \int_\Omega h_t(v_\varepsilon) \quad \text{as} \quad \varepsilon \to 0 \end{aligned}$$

with

$$h_t(s) \;:=\; \begin{cases} |s|^{p^*} & (|s| \geq t), \\ 0 & (\text{otherwise}). \end{cases}$$

By Fatou's lemma applied to the functions $|v_\varepsilon|^{p^*} - h_t(v_\varepsilon) \geq 0$, upper semi continuity of $h_t$, and the fact that $v_\varepsilon \to v$ a.e. the second term is asymptotically dominated by

$$\begin{aligned} \limsup_{\varepsilon \to 0} \int_\Omega h_t(v_\varepsilon) &\leq \int_\Omega \limsup_{\varepsilon \to 0} h_t(v_\varepsilon) \\ &\leq \int_\Omega h_t(v) \;=\; \int_{\{|v| \geq t\}} |v|^{p^*}. \end{aligned}$$

Letting first $\varepsilon \to 0$ and then $t \to \infty$ we obtain

$$F_0^- S^* \;\leq\; F^* \;=\; \lim_{\varepsilon \to 0} \varepsilon^{-p^*} \int_\Omega F(u_\varepsilon) \;\leq\; F_0^+ \lim_{\varepsilon \to 0} \int_\Omega |v_\varepsilon|^{p^*} \;\leq\; F_0^+ S^*$$

by Sobolev's inequality and normalization of $v_\varepsilon$. Since $F_0^- = F_0^+ > 0$ the sequence $(v_\varepsilon)$ is maximizing for $S^*$. By Sobolev's inequality $v_\varepsilon \to v$ in $L^{p^*}(\Omega)$. Thus $v$ is extremal for $S^*$ and must be of the form (3.6). In particular $\Omega = \mathbb{R}^n$ up to a set of p-capacity zero. This completes the proof of Theorem 4.13. □

In particular we found that if $F^* > F_0^+ S^*$ then every maximizing sequence for $F^*$ concentrates.

# Chapter 5

# Compactness Criteria

Let us recall the general compactness criterion (1.6) from the introduction. It says that if the global Sobolev constant is bigger than the maximal value of the local one, then an extremal function exists for the variational problem defining the global Sobolev constant. In this chapter we discuss two applications of this principle.

## 5.1 Extremal functions for Sobolev's inequality with anisotropic Dirichlet energy

As a first illustration of the general compactness criterion (1.6) we construct an anisotropic Dirichlet form for which the best Sobolev constant

$$S_A \quad := \quad \sup\left\{\int_\Omega |u|^{2^*} \;:\; u \in D^{1,2}(\Omega, A),\; \int_\Omega \nabla u \cdot A(\cdot)\nabla u \leq 1\right\}$$

is achieved on a bounded domain. This contrasts the isotropic case $A = I$ in which the best Sobolev constant is never achieved on bounded domains. It turns out that the compactness criterion $\sup S_A^{\mathrm{loc}} < S_A$ is satisfied as soon as the anisotropy of the matrix $A$ is strong enough. Otherwise the best Sobolev constant is either achieved or every maximizing sequence for $S_A$ concentrates at a minimum point of $\det A$. In the following let $A(x)$ be a symmetric, positive definite matrix depending continuously on $x \in \overline{\Omega}$.

**Lemma 5.1 (Anisotropic CCA)** *Under the above hypotheses on $A(x)$ the concentration compactness alternative for the critical power integrand (Lemma 4.2 and Theorem 4.3) extends to $D^{1,2}(\Omega, A)$ with the global Sobolev constant $S_A$ and the local constant*

$$S_A^{\mathrm{loc}}(x) \quad = \quad (\det A(x))^{-1/2} S^*.$$

*Proof.* The convexity argument (4.1) carries over to the present situation. We only need to prove the formula for the local Sobolev constant. By continuity of $A$ it suffices to consider the case of constant $A$. In this case

$$\int_\Omega |u|^{2^*} \leq (\det A)^{-1/2} S^* \left( \int_\Omega \nabla u \cdot A(\cdot) \nabla u \right)^{2^*/2}$$

for every $u \in D^{1,2}(\Omega, A)$. To see this we use the following scaling. After an orthogonal change of variables we can assume that $A = \operatorname{diag}(\lambda_1, \ldots, \lambda_n)$. For $u \in D^{1,2}(\Omega, A)$ define $v(Tx) := u(x)$ with the linear transformation $T = \operatorname{diag}(t_1, \ldots, t_n)$ with $t_k = \lambda_k^{-1/2}$. Then

$$\begin{aligned}
\int_\Omega \nabla u \cdot A(\cdot) \nabla u &= \int_\Omega \sum_k \lambda_k \left| \frac{\partial u}{\partial x_k} \right|^2 \\
&= \int_{T\Omega} \sum_k \lambda_k t_k^2 \left| \frac{\partial v}{\partial y_k} \right|^2 = \int_{T\Omega} |\nabla v|^2, \\
\int_\Omega |u|^{2^*} &= \int_{T\Omega} |v|^{2^*} \det T = \int_{T\Omega} |v|^{2^*} (\det A)^{-1/2}
\end{aligned}$$

and the assertion follows from the standard Sobolev inequality. □

By means of the same scaling one can prove the following principle of asymptotic directional equi-partition of energy for concentrating maximizing sequences. If a maximizing sequence for $(u_\varepsilon)$ for $S_A$ concentrates at $x$ then

$$(\nabla u_\varepsilon \cdot e)(A \nabla u_\varepsilon \cdot e) \stackrel{*}{\rightharpoonup} \frac{1}{n} \delta_x \text{ in } \mathcal{M}(\overline{\Omega})$$

for every unit vector $e \in \mathbb{R}^n$. In the isotropic case this means that $u_\varepsilon$ is asymptotically radial with respect to the concentration point. In the present application the general compactness criterion assumes the following form.

**Theorem 5.2 (Anisotropic compactness criterion)** *Under the above hypotheses on $A(x)$ we have:*

1. *If $\sup S_A^{\mathrm{loc}} < S_A$ then $S_A$ is attained.*

2. *If $S_A = \sup S_A^{\mathrm{loc}}$ then either $S_A$ is attained or every maximizing sequence for $S_A$ has a subsequence which concentrates at a minimum point of $\det A$.*

The following example shows that the compactness criterion (1.6) is satisfied provided the anisotropy of $A$ is strong enough. In this case $S_A$ is attained. For $n \geq 4$ let

$$\Omega = B_0^1 \times S^1 = \{x \in \mathbb{R}^n : |(x_1, \ldots, x_{n-1})| < 1\}$$

with $x$ and $y$ identified if $x_n - y_n \in \mathbf{Z}$ and $A = \operatorname{diag}(1, \ldots, 1, \lambda_n)$. We show that the compactness criterion is satisfied for $\lambda_n$ large enough. As a comparison function we choose $u(x) := v(x_1, \ldots, x_{n-1})$ with an extremal $v$ for

$$\sup \left\{ \int_{B_0^1} |v|^{\frac{2n}{n-2}} : v \in D^{1,2}(B_0^1),\ \|\nabla v\|_2 \le 1 \right\}$$

on the unit ball in $\mathbb{R}^{n-1}$. Such an extremal exists because the exponent $\frac{2n}{n-2}$ is subcritical in dimension $n-1$. Then $S_A \ge \int_{B_0^1} |v|^{\frac{2n}{n-2}}$ whereas

$$\sup S_A^{\mathrm{loc}} \;=\; \lambda_n^{-1/2} S^*$$

by Lemma 5.1. For large $\lambda_n$ the compactness criterion $\sup S_A^{\mathrm{loc}} < S_A$ is satisfied. Using an $x$-dependent matrix $A$ we can easily transform this example into one on a Euclidean domain, even a simply connected one. Similar examples can be constructed in three dimensions.

## 5.2 Extremal functions for Sobolev's inequality with conformal metrics

As a further illustration of the general compactness criterion (1.6) we study the question whether the best Sobolev constant is attained on domains whose metric is conformal to the Euclidean one. It turns out that the answer depends on the curvature. Nonexistence of extremal functions for negatively curved spaces is proved by means of a Pohozaev identity. Existence of extremal functions for large positively curved spaces is reduced to a result of Brézis and Nirenberg and a recent observation of Bandle. The problem of estimating the best Sobolev constants of manifolds has been addressed by Aubin [11]. Here we report on some of the results obtained with A. Brillard and C. Bandle [15] that fit into the present context.

Let $\Omega$ be a domain of dimension $n \ge 3$ endowed with a metric which is conformal to the Euclidean one. The conformality factor is denoted by $\sigma(x)$. It may be an arbitrary positive function on $\overline{\Omega}$ which is $C^2$. The corresponding gradient, and Laplace-Beltrami operator are

$$\begin{aligned} \nabla_\sigma u &:= \frac{\nabla u}{\sigma}, \\ L_\sigma u &:= \sigma^{-n} \operatorname{div}(\sigma^{n-2} \nabla u). \end{aligned}$$

Certain computations are more transparent when written in terms of $\nabla_\sigma$. The corresponding Sobolev constant is

$$S_\sigma(\Omega) \;:=\; \sup \left\{ \int_\Omega |u|^{2^*} \sigma^n : u \in D^{1,2}(\Omega, \sigma),\ \int_\Omega |\nabla u|^2 \sigma^{n-2} \le 1 \right\}.$$

For a constant conformality factor it is independent of $\sigma$ by homogeneity. Hence the local Sobolev constant for a continuous conformality factor is

$$S_\sigma^{\mathrm{loc}}(x) \quad = \quad S^*. \tag{5.1}$$

Moreover $S_\sigma(\Omega) \geq S^*$ and the general compactness criterion (1.6) assumes the following form.

**Theorem 5.3 (Compactness criterion for conformal metric)** *If $\sigma \in C(\overline{\Omega}, \mathbb{R}^+)$ and $S^* < S_\sigma(\Omega)$ then $S_\sigma(\Omega)$ is attained.*

The subsequent relation between the Dirichlet integral for the metric $\sigma\,|dx|$ and the Euclidean one will be used several times. The scalar curvature $\kappa$ of the metric $\sigma\,|dx|$ is defined by the relation

$$-\Delta\sigma^{\frac{n-2}{2}} \quad = \quad \frac{n-2}{4(n-1)}\kappa\sigma^{\frac{n+2}{2}}. \tag{5.2}$$

**Lemma 5.4** *For $u \in D^{1,2}(\Omega,\sigma)$ define $u_0 := u\sigma^{\frac{n-2}{2}}$. Then $u_0 \in D^{1,2}(\Omega)$ and*

$$\begin{aligned}
\int_\Omega |u|^{2^*}\sigma^n &= \int_\Omega |u_0|^{2^*}, \\
\int_\Omega |\nabla u|^2\sigma^{n-2} &= \int_\Omega |\nabla u_0|^2 - \frac{n-2}{4(n-1)}\int_\Omega |u_0|^2\kappa\sigma^2.
\end{aligned}$$

*Proof.* The gradient of $u$ is

$$\frac{\nabla u}{\sigma} \quad = \quad \frac{\nabla u_0}{\sigma^{n/2}} - \frac{n-2}{2}u_0\frac{\nabla\sigma}{\sigma^{\frac{n+2}{2}}}.$$

Therefore

$$|\nabla u|^2\sigma^{n-2} \quad = \quad |\nabla u_0|^2 - \frac{n-2}{2}\nabla|u_0|^2\cdot\frac{\nabla\sigma}{\sigma} + \left(\frac{n-2}{2}\right)^2|u_0|^2\left|\frac{\nabla\sigma}{\sigma}\right|^2.$$

Integration of the mixed term gives

$$\frac{n-2}{2}\int_\Omega |u_0|^2\operatorname{div}\frac{\nabla\sigma}{\sigma}$$

where

$$\frac{n-2}{2}\operatorname{div}\frac{\nabla\sigma}{\sigma} \quad = \quad \operatorname{div}\frac{\nabla\sigma^{\frac{n-2}{2}}}{\sigma^{\frac{n-2}{2}}} \quad = \quad -\frac{n-2}{4(n-1)}\kappa\sigma^2 - \left(\frac{n-2}{2}\right)^2\left|\frac{\nabla\sigma}{\sigma}\right|^2$$

as follows from (5.2). This yields the assertion. □

Based on a Pohozaev identity we obtain the following nonexistence result.

**Theorem 5.5** *If $\Omega$ is starshaped with respect to the origin and*

$$2\kappa\sigma^2 + \nabla(\kappa\sigma^2)\cdot x \quad \leq \quad 0 \ \ in \ \ \Omega$$

*then $S_\sigma(\Omega)$ is not achieved.*

*Proof.* If $u$ is an extremal for $S_\sigma(\Omega)$ then $u_0 = u\sigma^{\frac{n-2}{2}}$ solves

$$\begin{aligned} \Delta u_0 + \frac{n-2}{4(n-1)}\kappa\sigma^2 u_0 + \lambda u_0^{2^*-1} &= 0 \text{ in } \Omega, \\ u_0 &= 0 \text{ on } \partial\Omega \end{aligned}$$

as follows from Lemma 5.4. Testing this equation with $\nabla u_0 \cdot x$ and replacing $\int_\Omega |\nabla u_0|^2 - |u_0|^{2^*}$ by $\frac{n-2}{4(n-1)}\int_\Omega \kappa\sigma^2 |u_0|^2$ in the resulting Pohozaev identity yields

$$\frac{n-2}{4(n-1)}\int_\Omega \left(2\kappa\sigma^2 + \nabla(\kappa\sigma^2)\cdot x\right)|u_0|^2 \quad = \quad \int_{\partial\Omega} |\nabla u_0|^2 \, x\cdot\nu$$

where $\nu$ denotes the exterior unit normal of $\Omega$. By hypothesis the left hand side is negative, the right hand side positive. Thus $u_0 = 0$. □

The difference in the above identity to the standard Pohozaev identity (see e.g. Struwe [121]) is the second term on the left due to the x-dependence of the linear term. A more general Pohozaev identity with x-dependent coefficients can be found in Egnell [40, 43]. Now we specialize to the elliptic and the hyperbolic metric

$$\sigma_\pm(x) \quad := \quad \frac{2}{1 \pm |x|^2}.$$

An elliptic domain is a domain in $\mathbb{R}^n$ endowed with the metric $\sigma_+ |dx|$. A hyperbolic domain is a subset of the unit ball endowed with the metric $\sigma_- |dx|$. The scalar curvature of the elliptic and hyperbolic metric is $\kappa_\pm = \pm n(n-1)$ respectively. The nonexistence Theorem 5.5 applies to hyperbolic balls.

**Corollary 5.6** *The best Sobolev constant of an arbitrary hyperbolic domain is not achieved and $S_-(\Omega) = S^*$.*

*Proof.* For balls the first assertion follows from Theorem 5.5 because

$$2\kappa_-\sigma_-^2 + \nabla(\kappa_-\sigma_-^2)\cdot x \quad = \quad -2\sigma_-^2(1+\sigma_- |x|^2) \quad < \quad 0.$$

Using hyperbolic symmetrization with respect to the measure $\sigma_-^n dx$ (Baernstein [12]) the same follows for general domains. Indeed, assume $S_-(\Omega)$ has a nontrivial extremal $u$. Then

$$S^* \quad \leq \quad S_-(\Omega) \quad = \quad \int_{\Omega^*} |u^*|^{2^*} \sigma_-^n \quad \leq \quad S_-(\Omega^*).$$

If $S_-(\Omega^*) > S^*$ then $S_-(\Omega^*)$ would be attained by the compactness criterion (Theorem 5.3) in contradiction to Theorem 5.5. Therefore $S_-(\Omega^*) \leq S^*$. Thus the above inequalities imply that $u^*$ is a extremal for $S_-(\Omega^*)$ which is again impossible by Theorem 5.5. Hence $S_-(\Omega)$ is not attained. The second assertion is a consequence of the first one and the compactness criterion. □

The situation is different for elliptic domains.

**Corollary 5.7** *The best Sobolev constant $S_+(B_0^R)$ of a spherical cap is a non-decreasing function of $R$ which tends to $S^*$ as $R \to 0$ and to $\infty$ as $R \to \infty$. If $n \geq 4$ then*

$$S_+(B_0^R) \quad > \quad S^*$$

*for every $R > 0$. Consequently $S_+(\Omega)$ is always attained. In three dimensions*

$$\begin{aligned} S_+(B_0^R) \quad &> \quad S^* \quad \text{for} \quad R > 1, \\ S_+(B_0^R) \quad &= \quad S^* \quad \text{for} \quad R < 1. \end{aligned}$$

*Thus $S_+(\Omega)$ is attained if $\Omega$ contains a ball of radius 1.*

*Proof.* The admissible set of functions increases as the domain increases. This proves monotonicity. The limiting behaviour for small radii is a consequence of (5.1). Using the functions (3.6) as comparison functions with $s \to \infty$ we obtain the behaviour for large domains. By Lemma 5.4 we have

$$\begin{aligned} & S_+(B_0^R) \\ \geq & \sup_{u_0 \in D^{1,2}(B_0^R),\ \int_{B_0^R} |\nabla u_0|^2 - \frac{n(n-2)}{(1+R^2)^2} \int_\Omega |u_0|^2 \leq 1} \int_{B_0^R} |u_0|^{2^*}. \end{aligned}$$

The assertion follows from a result of Brézis and Nirenberg [27, Lemma 1.1]. For $n \geq 4$ the right hand side is always $> S^*$. In three dimensions we proceed as follows. A rough estimate using the capacity potential of $B_0^{R-1}$ with respect to $B_0^R$ as a comparison function yields

$$S_+(B_0^R) \ \geq \ \frac{c}{R} \ > \ S^*$$

for $R$ large enough. On the other hand

$$\begin{aligned} S^* \ &\leq \ S_+(B_0^R) \\ &\leq \ \sup \left\{ \int_{B_0^R} |u_0|^6 \ : \ u_0 \in D^{1,2}(B_0^R),\ \int_{B_0^R} |\nabla u_0|^2 - 3 \int_\Omega |u_0|^2 \leq 1 \right\} \end{aligned}$$

by Lemma 5.4. The result of Brézis and Nirenberg for the three dimensional case [27, Theorem 1.2] implies that the right hand side equals $S^*$ if $R \leq \pi/(2\sqrt{3})$. Bandle and Peletier [19] have recently shown that the critical radius for the three dimensional sphere is 1. This is the radius of the half sphere. □

Using harmonic transplantation and spherical symmetrization the statements for non-radial three dimensional domains can be sharpened [15]. Here we show numerically that the critical radius is 1. An extremal function for $S_+(B_0^R)$ solves

$$(5.3) \qquad \begin{aligned} u'' + \frac{n-1}{r} u' + n(n-2)\frac{u}{(1+r^2)^2} + u^{2^*-1} \quad &= \quad 0 \ \text{ on } \ (0, R), \\ u'(0) \quad &= \quad 0, \\ u(R) \quad &= \quad 0. \end{aligned}$$

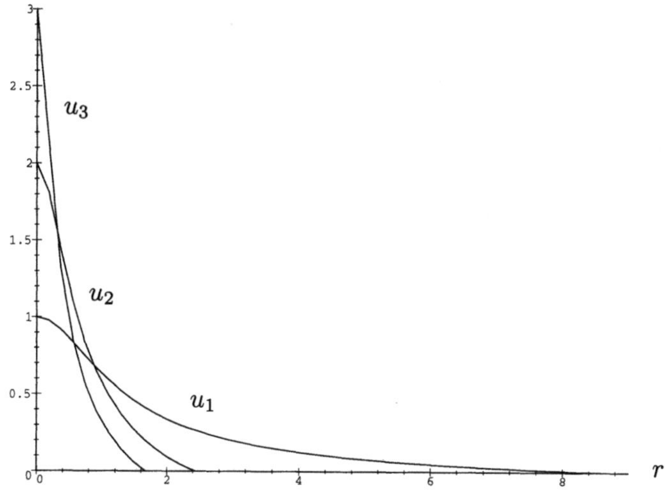


Figure 5.1: Solutions of (5.3) for $n = 3$ with $u(0) = 1, 2, 3$.

The unknown Lagrange multiplier has been eliminated by scaling. This does not affect the zeros. In the numerical procedure we fix $u(0)$ rather than $R$ such that a standard ODE solver can be used to approximate the solution $u = u_{u(0)}$. The radius $R$ is then determined as the first positive zero of $u_{u(0)}$. Figure 5.1 shows several solutions in three dimensions. As $u(0) \nearrow \infty$ we observe that $R \searrow 1$ as desired (Figure 5.2). Figure 5.3 shows that this phenomenon does not occur in higher dimensions.

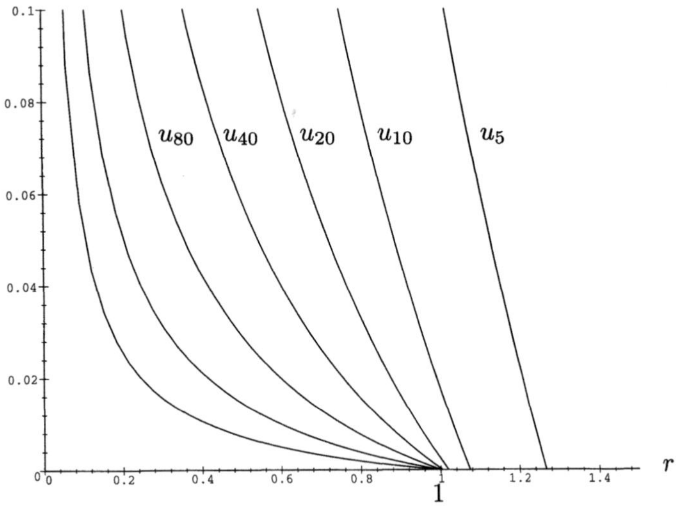


Figure 5.2: Solutions of (5.3) for $n = 3$ in the vicinity of 1 with $u(0) = 5, 10, 20, 40, 80, 160, 320$.

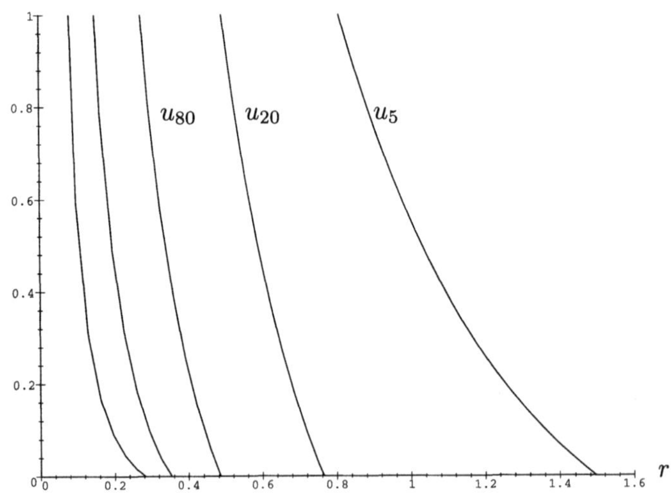


Figure 5.3: Solutions of (5.3) for $n = 4$ with $u(0) = 5, 20, 80, 320, 1280$.

# Chapter 6

# Entire Extremals

Extremal functions for the generalized Sobolev constant $F^*$ (Definition 3.1) are called *entire extremals.* Our main result concerning entire extremals is the second order decay estimate

$$w(r) \quad = \quad c\, r^{p-n}\,(1 + o(r))$$

in the $C^1$-sense together with an integral representation for the factor $c$ (Section 6.3). As a byproduct we solve a problem posed by Lions concerning radial symmetry of entire extremals. For smooth integrands an entire extremal is a ground state solution of

$$\begin{aligned} -\operatorname{div}\left(|\nabla w|^{p-2}\,\nabla w\right) \quad &= \quad \lambda f(w), \\ w(x) \quad &\to \quad 0 \;\text{ as }\; |x| \to \infty. \end{aligned} \tag{6.1}$$

Lions [90, Remark II.6] raised the question whether for non-symmetric integrands variational ground states are radial with constant sign. For $f \in C^{1,\alpha}$ with $f(0) = 0$ and $f'(0) < 0$ symmetry of positive solutions follows from the result of Gidas, Ni, and Nirenberg [64]. Symmetry for Lipschitz continuous nonlinearities is due to Kaper, Kwong, and Li [70]. We consider the zero mass case $f'(0) = 0$ without regularity assumptions for the nonlinearity. Thus the symmetry result of Gidas, Ni, and Nirenberg and its refinements do not apply. Nevertheless we give an affirmative answer to the question of Lions for the zero mass case. We show that for positive nonlinearities positive solutions are radially symmetric with respect to some point and strictly decreasing in radial direction. For general nonlinearities the same is true outside a compact set. This symmetry result is obtained as consequence of the decay estimates in combination with a characterization of the functions for which Schwarz symmetrization preserves the Dirichlet integral. In the semilinear case these results have been obtained with S. Müller [54].

A number of existence, symmetry, and decay rate results are available for the non-autonomous problem $-\Delta u = K(|x|)u^{\frac{n+2}{n-2}}$ in $\mathbb{R}^n$ covering the Matukuma equation as a special case (Ding and Ni [39], Li [83, 82], Li and Ni [84, 85], Bianchi and Egnell [23, 24]).

Uniqueness of radial ground states is an unsolved problem although it suffices to prove uniqueness of the number $w_\infty$ as defined in Lemma 6.4. Most results available deal with the positive mass case $f'(0) < 0$ or require a similar condition leading to exponential decay at infinity. Results of this type have been obtained by many authors including Kwong [76], Kwong and Zhang [78], Chen and Lin [33], Yanagida [132], Kwong and Li [77]. The most general uniqueness results (together with an up to date survey of the literature) are due to Franchi, Lanconelli, and Serrin [60]. They consider general divergence operators of the form $\operatorname{div}(A(|\nabla u|)\nabla u)$ inluding the p-Laplacian and the mean curvature operator.

The value of the Lagrange multiplier can be obtained from the observation that entire extremals are critical points of the functional

$$J(u) := \frac{\int_{\mathbb{R}^n} F(u)}{\|\nabla u\|_p^{p^*}}$$

defined on $D^{1,p}(\mathbb{R}^n) \setminus \{0\}$. This follows by scaling of the independent variable (3.1). The gradient of this functional is

$$\begin{aligned} J'(w)\eta &= \frac{d}{dt} J(w+t\eta)\bigg|_{t=0} \\ &= \int_{\mathbb{R}^n} f(w)\eta - p^* F^* \int_{\mathbb{R}^n} |\nabla w|^{p-2} \nabla w \cdot \nabla \eta \end{aligned}$$

by normalization of $w$ and $\int_{\mathbb{R}^n} F(w) = F^*$. Thus the Lagrange multiplier is given by

$$\lambda = \frac{1}{p^* F^*}. \tag{6.2}$$

## 6.1 Radial symmetry of entire extremals

If $w$ is an entire extremal then so is the symmetrized function $w^*$ as follows by scaling of the independent variable (3.1). Our symmetry result is based on the strict decay property of radial extremals (Theorem 6.5) as well as the subsequent characterization of the functions for which Schwarz symmetrization preserves the Dirichlet integral (see Kawohl [71] or Brothers and Ziemer [29]).

**Lemma 6.1** *If a positive function $u \in D^{1,p}(\mathbb{R}^n)$ with $p > 1$ satisfies $\|\nabla u^*\|_p = \|\nabla u\|_p$ then $u = u^*$ up to translation or $u$ has a plateau of positive volume which is below the top level.*

To begin with we show that every variational ground state is either strictly positive or strictly negative.

**Lemma 6.2 (Basic properties of entire extremals)** *Assume the growth condition* (1.3) *and let $w$ be an extremal for $F^*$. Then:*

1. *Either $w \geq 0$ or $w \leq 0$. For definiteness we assume $w \geq 0$.*

2. *There is a ball $B_{x_0}^R$ (we assume $x_0 = 0$) such that $w = w^*$ outside this ball. In this range $w$ is strictly decreasing in radial direction. In particular $w > 0$ in $\mathbb{R}^n$.*

*Proof.* If $w$ changes sign we split $w = w_+ + w_-$ with $\|\nabla w_+\|_p < 1$, $S_+ := \int_{\mathbb{R}^n} F(w_+) < F^*$, and similarly for the negative part. Both functions $w_\pm$ belong to $D^{1,2}(\mathbb{R}^n)$. Normalization by scaling of the independent variable (3.2) leads to the contradiction

$$\begin{aligned} F^* &= S_+ + S_- = \|\nabla w_+\|_p^{p^*} \int_{\mathbb{R}^n} F(w_+^{s_+}) + \|\nabla w_-\|_p^{p^*} \int_{\mathbb{R}^n} F(w_-^{s_-}) \\ &\leq F^* \left( \left( \int_{\mathbb{R}^n} |\nabla w_+|^p \right)^{p^*/p} + \left( \int_{\mathbb{R}^n} |\nabla w_-|^p \right)^{p^*/p} \right) < F^* \end{aligned}$$

by strict convexity of the function $t \mapsto t^{p^*/p}$ on $\mathbb{R}^+$. Thus $w \geq 0$ or $w \leq 0$.

By Theorem 6.5 below the symmetrized function $w^*$ is strictly decreasing outside some ball $B_0^R = \{w^* > t\}$. We show that after translation $w = w^*$ in $\mathbb{R}^n \setminus B_0^R$. Assume the contrary. Application of Lemma 6.1 to the function $\min(w, t)$ shows that the function $w^*$ has a doubly connected plateau $B^{R_2} \setminus B^{R_1}$ with $R_2 > R_1 > R$ in contradiction to the fact that $w^*$ is strictly decreasing outside $B_0^R$. □

The example of Section 3.3 (volume integrand) shows that extremal functions may have flat parts on the top level.

## 6.2 Euler Lagrange equation (variation of the independent variable)

The integrand $F$ may be discontinuous. Thus we cannot use the usual Euler Lagrange equation. Instead we derive an Euler Lagrange equation corresponding to variations of the independent variable. It involves the integrand $F$ but no derivatives of $F$. If $w$ is a radial function we write $r := |x|$ and $w(r) := w(|x|)$. A jump, i.e. the difference between right and left limit of a function is denoted by square brackets.

**Lemma 6.3 (Euler equation by variation of independent variable)** *Every radial extremal for $F^*$ satisfies*

$$(r^q (\gamma |w'|^p + F(w)))' = q r^{q-1} F(w) \tag{6.3}$$

*in the sense of distributions with*

$$\begin{aligned} q &:= p\frac{n-1}{p-1} > n, \\ \gamma &:= \frac{n(p-1)F^*}{n-p}. \end{aligned}$$

*In particular the jump condition* $\gamma\left[|w'|^p\right] = -[F(w)]$ *holds at the discontinuities of* $F$.

*Proof.* Consider the variations

$$\eta_s(r) := r + s\eta(r), \qquad w_s := w \circ \eta_s^{-1}$$

of the independent variable with $s$ so small that $\eta_s' > 0$. Then

$$\begin{aligned} \frac{d}{ds}\int_0^\infty r^{n-1}F(w_s)\bigg|_{s=0} &= \frac{d}{ds}\int_0^\infty F(w)\left(\frac{1}{n}\eta_s^n\right)'\bigg|_{s=0} \\ &= \int_0^\infty F(w)(r^{n-1}\eta)' \\ &= \int_0^\infty r^{n-1}F(w)\eta' + (n-1)\int_0^\infty r^{n-2}F(w)\eta. \end{aligned}$$

Similarly

$$\begin{aligned} &\frac{d}{ds}\int_0^\infty r^{n-1}\left|w_s'\right|^p\bigg|_{s=0} \\ =\ &\frac{d}{ds}\int_0^\infty \eta_s^{n-1}\left|w'\right|^p(\eta_s')^{1-p}\bigg|_{s=0} \\ =\ &-(p-1)\int_0^\infty r^{n-1}\left|w'\right|^p\eta' + (n-1)\int_0^\infty r^{n-2}\left|w'\right|^p\eta. \end{aligned}$$

Inserting these terms into

$$\frac{d}{ds}J(w_s)\bigg|_{s=0} = \frac{d}{ds}\left(\int_0^\infty r^{n-1}F(w_s) - \frac{n}{n-p}F^*\int_0^\infty r^{n-1}\left|\nabla w_s\right|^p\right)\bigg|_{s=0}$$

with $\eta(r) = r\psi(r)$ yields (6.3) with the test function $\psi$. □

For smooth integrands (6.3) is equivalent to

$$(p-1)\left|w'\right|^{p-2}w'' + \frac{n-1}{r}\left|w'\right|^{p-2}w' + \lambda f(w) = 0$$

at all points where $w'$ is defined and nonzero. This is the radial part of the usual Euler Lagrange equation (6.1).

## 6.3 Second order decay estimate for entire extremals

We first assume that $F$ is non-decreasing on $\mathbb{R}^+$ and non-increasing on $\mathbb{R}^-$. For smooth integrands this corresponds to $f \geq 0$ on $\mathbb{R}^+$ and $f \leq 0$ on $\mathbb{R}^-$ respectively. This hypothesis is satisfied in many applications. It implies that every positive extremal is super p-harmonic in the sense of distributions. On every subdomain a super p-harmonic function is pointwise bigger or equal to the p-harmonic function with the same boundary values. The properties listed below are consequences of this fact.

**Lemma 6.4 (Monotone integrands)** *Assume the growth condition* (1.3) *and let $w$ be a positive extremal for $F^*$. If $F$ is non-decreasing on $\mathbb{R}^+$ then:*

1. *After translation we have $w = w^*$ (we assume that $w$ is radial with respect to the origin).*
2. *The function $r \mapsto w(r)$ is strictly decreasing on $\{w < w(0)\}$.*
3. *The function $r \mapsto w(r)/K(r)$ is non-decreasing on $\mathbb{R}^+$.*
4. *The function $r \mapsto w'(r)/K'(r)$ is non-decreasing on $\mathbb{R}^+$. In particular every kink satisfies $[w'] \leq 0$.*
5. *The limit*

$$w_\infty \quad := \quad \lim_{r\to\infty} \frac{w(r)}{K(r)} \quad = \quad \lim_{r\to\infty} \frac{w'(r)}{K'(r)}$$

*exists and $w_\infty > 0$. In particular $w(r) \leq cr^{-\frac{n-p}{p-1}}$ and $F(w(r)) \leq cr^{-\frac{np}{p-1}}$.*

*Proof.* The symmetrized function $w^*$ is also extremal. By weak super p-harmonicity of $w^*$ outside the ball $B_0^r$ we have

$$w^*(R) \quad > \quad \frac{w^*(r)}{K(r)} K(R)$$

for every $R \geq r$. Thus $r \mapsto w(r)/K(r)$ is non-decreasing. Moreover $w^*$ is strictly decreasing below the top level and strictly positive. Lemma 6.1 implies $w = w^*$ up to translation. Weak super p-harmonicity of $w$ implies that for every triple $r \leq s \leq R$ we have

$$w(s) \quad \geq \quad w(R) + \frac{K(s) - K(R)}{K(r) - K(R)}(w(r) - w(R)).$$

Therefore

$$w'(r) \quad \geq \quad K'(r)\frac{w(r) - w(R)}{K(r) - K(R)}, \qquad w'(R) \quad \leq \quad K'(R)\frac{w(r) - w(R)}{K(r) - K(R)}$$

which implies $w'(r)/K'(r) \leq w'(R)/K'(R)$. Thus the limit $w_\infty$ exists in $(0, \infty]$. We are left to show that it is finite. This is trivial if $F = 0$ near 0. For general integrands we refer to Theorem 6.5 below. □

In the semilinear case a symmetry result in this direction for Lipschitz continuous $f$ was given by Kaper, Kwong, and Li [70]. Now we return to general integrands which are not necessarily monotone.

**Theorem 6.5 (Decay rate of entire extremals)** *Assume the growth condition* (1.3) *and let $w = w^*$ be a positive radial extremal for $F^*$. Then*

$$\begin{aligned} w_\infty K(r)\left(1 - O(r^{-\frac{p}{p-1}})\right) &\le w(r) \le w_\infty K(r), \\ w_\infty K'(r) &\le w'(r) \le w_\infty K'(r)\left(1 - O(r^{-\frac{p}{p-1}})\right), \\ \int_{\mathbb{R}^n \setminus B_0^R} |\nabla w|^p &= w_\infty^p K(R)\left(1 + O(R^{-\frac{p}{p-1}})\right) \end{aligned}$$

*as $r,\ R \to \infty$ where*

$$w_\infty{}^p = \frac{p(n-1)}{n(p-1)F^*} \int_{\mathbb{R}^n} \frac{F(w)}{K(|\cdot|)}.$$

*In particular $w'(r) < 0$ for $r > R \ge 0$, $w(r) \le c\, r^{-\frac{n-p}{p-1}}$, $F(w(r)) \le c\, r^{-\frac{np}{p-1}}$, and*

$$\int_{\mathbb{R}^n \setminus B_0^R} F(w) \le c R^{-\frac{n}{p-1}}$$

*for every $R > 0$.*

*Proof.* In terms of the auxiliary functions

$$\begin{aligned} a(r) &:= r^n \left(\gamma\, |w'(r)|^p + F(w(r))\right), \\ b(r) &:= r^n F(w(r)) \end{aligned}$$

equation (6.3) assumes the form

$$(r^{q-n} a(r))' = q\, r^{q-n-1} b(r). \tag{6.4}$$

Recall that $q = p\frac{n-1}{p-1}$ and $\gamma = \frac{n(p-1)F^*}{n-p}$. Equivalently we can say that

$$a'(r) = r^{n-1}\left(nF(w) - \frac{n-p}{p-1}\gamma\, |w'|^p\right).$$

The right hand side is in $L^1(\mathbb{R}^+)$. Thus the limit $a(\infty)$ exists. Together with

$$\int_0^\infty \frac{a(r)}{r} = \int_0^\infty r^{n-1}\left(\gamma\, |w'|^p + F(w)\right) < \infty$$

we conclude that $a(\infty) = 0$. Also the decreasing envelope

$$A(r) := \sup_{R \ge r} a(r)$$

tends to 0. In terms of this function we can estimate

$$w(R) \;=\; \int_\infty^R w' \;\leq\; c\int_R^\infty \left(\frac{a(r)}{r^n}\right)^{1/p} \;\leq\; c\,A(R)^{1/p} R^{-\frac{n-p}{p}}$$

and therefore

$$b(r) \;\leq\; c r^n w(r)^{p^*} \;\leq\; c\,A(r)^{p^*/p} \tag{6.5}$$

by the growth condition (1.3). Integration of (6.4) over the interval $(r, R)$ yields

$$a(R) - \left(\frac{r}{R}\right)^{q-n} a(r) \;\leq\; c\left(1 - \left(\frac{r}{R}\right)^{q-n}\right) A(r)^{p^*/p} \;\leq\; c\,A(r)^{p^*/p}.$$

After taking the supremum over all $R' \geq R$ we obtain

$$A(R) \;\leq\; \left(\left(\frac{r}{R}\right)^{q-n} + A(r)^{\frac{p}{n-p}}\right) A(r) \quad \text{for every} \quad R > r > 0. \tag{6.6}$$

For every positive exponent $\beta < q-n$ we can find $r > 0$ such that

$$\left(\frac{1}{2}\right)^{(q-n)} + A(r)^{\frac{p}{q-n}} \;\leq\; \left(\frac{1}{2}\right)^{\beta} \;<\; 1$$

because $A(r) \to 0$ as $r \to \infty$. By (6.6) the same relation holds for all $R \geq r$. Iterated application of (6.6) with $R = 2r$ and monotonicity of $A$ yields

$$A(r) \;\leq\; cr^{-\beta}, \quad b(r) \;\leq\; cr^{-\beta p^*/p}$$

for every $r$ by (6.5). Choosing $\beta > \frac{p}{p^*}(q-n)$ the right hand side of (6.4) is integrable. In particular $a(r) \leq c\,r^{n-q}$. Another application of (6.5) yields

$$b(r) \;\leq\; cr^{-n\frac{q-n}{n-p}} \;=\; cr^{-\frac{n}{p-1}}, \quad F(w(r)) \;\leq\; cr^{-\frac{np}{p-1}}$$

and $r^{q-n-1}b(r) \to 0$ as $r \to \infty$. Integration of (6.4) together with $a(0) = 0$ yields

$$r^q\left(\gamma\,|w'|^p + F(w)\right) \;\leq\; c_1 \;\; \text{with}$$
$$c_1 \;:=\; q\int_0^\infty r^{q-1}F(w) \;=\; \frac{q}{|S^{n-1}|}\int_{\mathbb{R}^n} |x|^{q-n}\, F(w)$$

for every $r$ with equality at infinity. Thus we obtain

$$|w'|^p \;\leq\; \frac{c_1}{\gamma} r^{-q} \;=\; w_\infty^p\, |K'(r)|^p$$

which yields the integral representation for $w_\infty$ and

$$w(r) \;\leq\; \int_r^\infty \gamma\,|w'| \;\leq\; \left(\frac{c_1}{\gamma}\right)^{1/p} r^{1-\frac{q}{p}} \;=\; w_\infty K(r).$$

Again by (6.4)

$$R^q\left(\gamma\left|w'\right|^p+F(w)\right) = c_1-q\int_R^\infty r^{q-n-1}b(r) = c_1-O(R^{-\frac{p}{p-1}})$$

since $b(r)\le cr^{-\frac{n}{p-1}}$. Together with $F(w(r))\le cr^{-\frac{np}{p-1}}$ we obtain $w'(r) = w_\infty K'(r)\left(1+O(r^{-\frac{p}{p-1}})\right)$ and the lower bound for $w$. □

The approximation formulas of Theorem 6.5 are best possible as can be seen from the extremal

$$w(r) = \frac{c}{\left(1+r^{\frac{p}{p-1}}\right)^{\frac{n-p}{p}}} = w_\infty K(r)\left(1+O(r^{-\frac{p}{p-1}})\right)$$

for $S^*$. In the semilinear case $0\le f(t)\le t^q$ with $q>\frac{n+2}{n-p}$ Egnell [44, Proposition C] found

$$w_\infty = \lambda\int_{\mathbb{R}^n} f(w).$$

Formally this follows by integration of (6.1) over a large ball. Egnell's approach is not applicable to the p-Laplacian because he uses Kelvin transformation.

For $f$ vanishing for small values our result intersects with that of Li and Ni [86] saying that positive solutions of $-\Delta u=f(u)$ on $\mathbb{R}^n$ with $f'(t)\le 0$ for $0<t<t_0$ are radial and strictly decreasing in radial direction. We expect that entire extremals are unique up to sign and translation. Most uniqueness results available today deal with the semilinear positive mass case $f(0)'<0$ or require a similar condition leading to exponential decay of ground state solutions. Results of this type are due to Kwong [76], Kwong and Zhang [78], Chen and Lin [33]. The problem $-\Delta w = w^{q-1}$ with subconformal $q$ has two radially decaying solutions. One of them decays like $r^{-(n-2)}$, the other like $r^{-\frac{q}{q-2}}$ (Bandle and Marcus [18]). This shows that the growth condition is necessary for the decay rate $K(r)$ and for uniqueness.

# Chapter 7

# Concentration and Limit Shape of Low Energy Extremals

In this chapter we show that low energy extremals concentrate at a single point and we analyze their local behaviour near the concentration point. The corresponding semilinear results have been obtained with S. Müller [55]. In the smooth case additional information can be derived from the Euler Lagrange equation. For the supremum in the variational problem (1.2) we use the following abbreviation.

**Definition 7.1**

$$F_\Omega^\varepsilon \quad := \quad \sup\left\{\int_\Omega F(u) \;:\; u \in D^{1,p}(\Omega),\ \|\nabla u\|_p \leq \varepsilon\right\}.$$

It is related to the generalized Sobolev constant in the following way.

**Proposition 7.2** *If the growth condition* (1.3) *holds then*

$$\begin{aligned}
\varepsilon^{-p^*} F_\Omega^\varepsilon \quad &\leq \quad F^* \quad \textit{for every } \varepsilon > 0 \quad , \\
\lim_{\varepsilon\to 0} \varepsilon^{-p^*} F_\Omega^\varepsilon \quad &= \quad F^*, \\
\max(F_0^-, F_\infty^-) \quad &\leq \quad \frac{F^*}{S^*}.
\end{aligned}$$

*Proof.* The first claim is a reformulation of the generalized Sobolev inequality (Lemma 3.2). For the second one fix $\delta > 0$, $x_0 \in \Omega$, and a candidate $w$ for the definition of $F^*$ satisfying $\int_{\mathbb{R}^n} F(w) \geq F^* - \delta$. For $r$ large enough $\int_{B_0^r} F(w) \geq F^* - 2\delta$. By Lemma 3.3 there is a cut-off function $\eta$ with $\eta = 1$ in $B_0^r$ vanishing outside $B_0^R$ such that $\int_{B_0^R} |\nabla(\eta w)|^p \leq 1 + \delta$. For $\varepsilon$ sufficiently small

$$u(x) \quad := \quad (\eta w)\left(\left(\frac{1+\delta}{\varepsilon^p}\right)^{\frac{1}{n-p}}(x - x_0)\right)$$

has its support in $\Omega$ and $\|\nabla u\|_p \leq \varepsilon$ by (3.1). Thus

$$\begin{aligned}\varepsilon^{-p^*} F_\Omega^\varepsilon &\geq \varepsilon^{-p^*} \int_\Omega F(u) \\ &= (1+\delta)^{-\frac{n}{n-p}} \int_{B_0^R} F(\eta w) \geq (1+\delta)^{-\frac{n}{n-p}} (F^* - 2\delta)\end{aligned}$$

The assertion follows as $\delta \to 0$. To prove the last inequality we choose an extremal $w$ for $S^*$. It is of the form (3.6). The functions $s^{-\frac{n-p}{p}} w^s$ are admissible for the definition of $F^*$. They tend uniformly to 0 as $s \to \infty$. Therefore

$$\begin{aligned}F^* &\geq \int_{\mathbb{R}^n} F\left(s^{-\frac{n-p}{p}} w^s\right) \\ &= s^n \int_{\mathbb{R}^n} F\left(s^{-\frac{n-p}{p}} w\right) \geq (F_0^- + o(1)) \int_{\mathbb{R}^n} |w|^{p^*}.\end{aligned}$$

The inequality for $F_0^-$ follows as $s \to \infty$. Letting $s \to 0$ we obtain the inequality for $F_\infty^-$ in a similar way. □

Existence of an extremal function $u_\varepsilon$ for $F_\Omega^\varepsilon$ follows from the general hypotheses (1.3). In fact, consider a maximizing sequence $(u_k)$. A subsequence converges weakly in $D^{1,p}(\Omega)$, strongly in $L^p(\Omega)$, and therefore a.e. to a function $u_\varepsilon$. By Fatou's lemma and upper semi continuity of the integrand the weak limit is an extremal.

## 7.1 Concentration of low energy extremals

Now we are in position to prove the first of the three main results on low energy limits.

**Theorem 7.3 (Concentration theorem)** *Assume the growth condition* (1.3) *and one of the hypotheses*

1. $F_0^+ < F^*/S^*$,
2. $F_0^+ = F_0^-$ *and* $\Omega \neq \mathbb{R}^n$ *up to a set of positive p-capacity,*
3. $\Omega$ *has finite volume.*

*For every* $\varepsilon > 0$ *let* $u_\varepsilon$ *be an extremal function for* $F_\Omega^\varepsilon$. *Then a subsequence of* $(u_\varepsilon)$ *concentrates at some point* $x_0 \in \overline{\Omega}$, *i.e.*

$$\frac{|\nabla u_\varepsilon|^p}{\varepsilon^p} \stackrel{*}{\rightharpoonup} \delta_{x_0} \text{ in } \mathcal{M}(\overline{\Omega}), \qquad \varepsilon^{-p^*} F(u_\varepsilon) \stackrel{*}{\rightharpoonup} F^* \delta_{x_0} \text{ in } \mathcal{M}(\overline{\Omega}).$$

*Proof.* A subsequence of $(u_\varepsilon)$ satisfies the hypotheses of the concentration compactness alternative for low energy extremals (Theorem 4.13). By Proposition 7.2 we have $\nu(\overline{\Omega}) = F^*$ and it suffices to exclude compactness. Suppose on the contrary that compactness occurs. Then $v_\varepsilon = u_\varepsilon/\varepsilon \to v \neq 0$ in $L^{p^*}(\Omega)$ and a.e. and $F^* \le F_0^+ S^*$ according to Theorem 4.13. Thus compactness is excluded if $F_0^+ < F^*/S^*$. We may assume $F_0^+ > 0$ since otherwise $F^* = 0$. The assertion in the case $F_0^- = F_0^+$ follows from the last statement of Theorem 4.13. For domains of finite volume we first consider the case $u_\varepsilon \ge 0$. Fix $t > 0$ and let $r_\varepsilon(t)$ be the radius of the ball $\{v_\varepsilon{}^* > t\}$. Define $\overline{v}_\varepsilon = v_\varepsilon^*$ on $\{v_\varepsilon{}^* > t\}$ and extend it to all of $\mathbb{R}^n$ by the p-harmonic function vanishing at infinity. By the formula for the capacity of balls (2.7) we have

$$\int_{\mathbb{R}^n} |\nabla \overline{v}_\varepsilon|^p \;=\; \int_{\{v_\varepsilon^* > t\}} |\nabla v_\varepsilon^*|^p + t^p K(r_\varepsilon(t))^{1-p}.$$

Since $\|\nabla v_\varepsilon^*\|_p \le \|\nabla v_\varepsilon\|_p \le 1$ we can estimate the first part of the p-Dirichlet integral by

$$\int_{\{v_\varepsilon^* > t\}} |\nabla v_\varepsilon^*|^p \;\le\; 1 - \int_{\{v_\varepsilon^* \le t\}} |\nabla v_\varepsilon^*|^p \;\le\; 1 - t^p \left(K(r_\varepsilon(t)) - K(r_\varepsilon(0))\right)^{1-p}.$$

The radius $R_\Omega$ of the ball $\Omega^*$ is finite by hypothesis and $r(t) := \liminf_{\varepsilon\to 0} r_\varepsilon(t)$. Then $r(0) \le R_\Omega$. For small $t$ we have $r(t) \ge c > 0$ because $v \neq 0$ and $|\{v > t\}| \le \liminf_{\varepsilon\to 0} |\{v_\varepsilon \ge t\}|$. Combining the above estimates yields

$$\begin{aligned}\limsup_{\varepsilon\to 0} \int_{\mathbb{R}^n} |\nabla \overline{v}_\varepsilon|^p \;&\le\; 1 - t^p \left( (K(r(t)) - K(r(0)))^{1-p} - K(r(t))^{1-p} \right) \\ &\le\; 1 - c\,t^p r(t)^{n-p}.\end{aligned}$$

The functional changes according to

$$\begin{aligned}\varepsilon^{-p^*} \int_{\mathbb{R}^n} F(\varepsilon \overline{v}_\varepsilon) \;&\ge\; \varepsilon^{-p^*} \int_{\{v_\varepsilon^* > t\}} F(\varepsilon v_\varepsilon^*) \\ &=\; \varepsilon^{-p^*} \int_\Omega F(u_\varepsilon) - \varepsilon^{-p^*} \int_{\{v_\varepsilon \le t\}} F(\varepsilon v_\varepsilon) \\ &\ge\; F^* + o(\varepsilon^0) - c \int_{\{v_\varepsilon \le t\}} |v_\varepsilon|^{p^*} \;\ge\; F^* - c\,|\Omega|\, t^{p^*}\end{aligned}$$

for small $\varepsilon$. Using the generalized Sobolev inequality (Lemma 3.2) we conclude

$$\begin{aligned}F^* - C\,t^{p^*} \;&\le\; \limsup_{\varepsilon\to 0} \varepsilon^{-p^*} \int_{\mathbb{R}^n} F(\varepsilon \overline{v}_\varepsilon) \\ &\le\; F^* (1 - c\,t^p)^{p^*/p} \;\le\; F^* - c\,t^p.\end{aligned}$$

For small $t$ this is a contradiction because $p < p^*$. If $v_\varepsilon$ changes sign we symmetrize the positive and the negative part separately using the relations

$$\begin{aligned}\int_{\mathbb{R}^n} |\nabla u_+^*|^p + \int_{\mathbb{R}^n} |\nabla u_-^*|^p \;&\le\; \int_{\mathbb{R}^n} |\nabla u|^p, \\ \int_{\mathbb{R}^n} F(u_+^*) + \int_{\mathbb{R}^n} F(u_-^*) \;&=\; \int_{\mathbb{R}^n} F(u).\end{aligned}$$

□

We expect that the conditions conditions (1.3) and $\Omega \neq \mathbb{R}^n$ up to a set of p-capacity zero already imply concentration. The latter condition is not obsolete. For the critical power integrand on $\mathbb{R}^n$ the functions $u_\varepsilon = \varepsilon w$ with an extremal $w$ for $S^*$ as in (3.6) are extremals for $S^*_\varepsilon(\mathbb{R}^n)$ but

$$\frac{|\nabla u_\varepsilon|^p}{\varepsilon^p} \quad = \quad |\nabla w|^p$$

does not tend to a Dirac measure. The concentration theorem (Theorem 7.3) applies to the volume integrand (Section 3.3) and to the plasma integrand (Section 3.4). In both examples we have $F_0 = F_\infty = 0$. If $\Omega$ is not all of $\mathbb{R}^n$ we obtain concentration even for the critical plasma problem.

## 7.2 Limit shape of low energy extremals

From the previous section we know that low energy extremals concentrate at a single point. Next we analyze their asymptotic behaviour near the concentration point. We will see that after a suitable scaling of the independent variable they tend to an entire extremal. This is the second of the announced main results.

**Theorem 7.4 (Limit shape of low energy extremals)** *If the growth condition* (1.3) *holds and*

$$\max(F_0^+, F_\infty^+) \quad < \quad \frac{F^*}{S^*} \tag{7.1}$$

*then there are points $x_\varepsilon \to x_0$ with $x_0$ given by the concentration theorem (Theorem* 7.3*) such that the rescaled functions*

$$\begin{aligned} w_\varepsilon(y) \quad &:= \quad u_\varepsilon\left(x_\varepsilon + \varepsilon^{\frac{p}{n-p}} y\right) \quad \textit{supported on} \\ \Omega_\varepsilon \quad &:= \quad \varepsilon^{-\frac{p}{n-p}}(\Omega - x_\varepsilon) \end{aligned}$$

*tend to an extremal for $F^*$, i.e. $w_\varepsilon \to w$ in $D^{1,p}(\mathbb{R}^n)$, $\|\nabla w\|_p = 1$, and $\int_{\mathbb{R}^n} F(w) = F^*$.*

*Proof.* Note that $\|\nabla w_\varepsilon\|_p \leq 1$ and $\int_{\mathbb{R}^n} F(w_\varepsilon) = \varepsilon^{-p^*} \int_\Omega F(u_\varepsilon) \to F^*$. We apply the *compactness-splitting-vanishing alternative* of Lions [89, p. 115] to the measures

$$\sigma_\varepsilon \quad := \quad F(\widetilde{w}_\varepsilon) \text{ with } \widetilde{w}_\varepsilon(y) \quad := \quad u_\varepsilon\left(\varepsilon^{\frac{p}{n-p}} y\right).$$

**Lemma 7.5 (Compactness-splitting-vanishing alternative)** *Every sequence $(\sigma_\varepsilon)$ of positive Borel measures on $\mathbb{R}^n$ with*

$$\sigma_\varepsilon(\mathbb{R}^n) \quad \to \quad S > 0 \quad \textit{as} \quad \varepsilon \to 0$$

*has a subsequence for which one of the following statements holds true.*

1. *Compactness: the measures $(\sigma_\varepsilon)$ converge up to translation, i.e. there is a sequence of centers $(x_\varepsilon)$ in $\mathbb{R}^n$ such that for every $\delta > 0$ a radius $R$ exists for which*

$$\sigma_\varepsilon(B^R_{x_\varepsilon}) \;\geq\; S - \delta$$

*for every $\varepsilon$.*

2. *Splitting: The measures $(\sigma_\varepsilon)$ separate into two distant pieces, i.e. there are constants $S_1$, $S_2 > 0$ with $S_1 + S_2 = S$, such that for every $\delta > 0$ a radius $r$ and centers $(x_\varepsilon)$ exist, such that for every $R > r$*

$$\begin{aligned} \sigma_\varepsilon(B^r_{x_\varepsilon}) &\geq S_1 - \delta, \\ \sigma_\varepsilon(\mathbb{R}^n \setminus B^R_{x_\varepsilon}) &\geq S_2 - \delta \end{aligned}$$

*for $\varepsilon$ sufficiently small.*

3. *Vanishing: The measures $(\sigma_\varepsilon)$ smear out in the sense that for every radius $R$ one has*

$$\lim_{\varepsilon \to 0} \sup_{x \in \mathbb{R}^n} \sigma_\varepsilon(B^R_x) \;=\; 0.$$

The strategy is as follows. We first prove compactness in the above sense for the measures $\sigma_\varepsilon$ by application of the compactness-splitting-vanishing alternative. This requires exclusion of splitting and vanishing. Then we apply the generalized concentration compactness alternative to obtain strong convergence which requires exclusion of concentration. The condition $F_0^+ < F^*/S^*$ prevents the sequence from vanishing, the condition $F_\infty^+ < F^*/S^*$ from concentration. In the following we can assume $x_\varepsilon = 0$.

**Step 7.6** *Exclusion of splitting.*

*Proof.* Suppose for a contradiction that there are $S_1$, $S_2 > 0$ with $S_1 + S_2 = F^*$, such that for every $\delta > 0$ a radius $r$ exists, such that

$$\begin{aligned} S_1 - \delta &\leq \int_{B_0^r} F(\widetilde{w}_\varepsilon), \\ S_2 - \delta &\leq \int_{\mathbb{R}^n \setminus B_0^R} F(\widetilde{w}_\varepsilon) \end{aligned}$$

for all $R > r$ and $\varepsilon$ sufficiently small. For every $\delta' > 0$ the local generalized Sobolev inequality (Corollary 3.4) provides two radii $R$, $r'$ such that $R > r' > r$ and

$$\begin{aligned} S_1 - \delta &\leq \int_{B_0^r} F(\widetilde{w}_\varepsilon) \;\leq\; F^* \left( \int_{B_0^{r'}} |\nabla \widetilde{w}_\varepsilon|^p + \delta' \right)^{p^*/p}, \\ S_2 - \delta &\leq \int_{\mathbb{R}^n \setminus B_0^R} F(\widetilde{w}_\varepsilon) \;\leq\; F^* \left( \int_{\mathbb{R}^n \setminus B_0^{r'}} |\nabla \widetilde{w}_\varepsilon|^p + \delta' \right)^{p^*/p}. \end{aligned}$$

Adding these inequalities we obtain

$$\left(\frac{S_1-\delta}{F^*}\right)^{\frac{n-p}{n}} + \left(\frac{S_2-\delta}{F^*}\right)^{\frac{n-p}{n}} \leq 1+2\delta'.$$

For $\delta'$ and $\delta$ sufficiently small this is a contradiction since $S_1 > 0$ and $S_2 = F^* - S_1 > 0$. □

In the next step we use the following variant of the standard Poincaré-Sobolev inequality. The corresponding variant of Poincaré's inequality can be found in Morrey's book [96, Theorem 3.6.5].

**Lemma 7.7** *Suppose $Q$ is a connected domain of finite volume with Lipschitz boundary and let $\theta > 0$. Then there is a constant $c$ such that*

$$\|w\|_{p^*} \leq c\|\nabla w\|_p$$

*for every $w \in H^{1,p}(Q)$ with $|\{w=0\}| \geq \theta$.*

*Proof.* Suppose the assertion was false. Then there exists a sequence $(w_k)$ such that $\|w_k\|_{p^*} = 1$, $|\{w_k = 0\}| \geq \theta$ and $\|\nabla w_k\|_p \to 0$. By Rellich's theorem $w_k \to w$ in $L^p(Q)$ for a subsequence. Hence $w_k \to w$ in $H^1(Q)$ and $L^{p^*}(Q)$. Moreover $\nabla w = 0$ a.e. By connectedness of $Q$ the limit function is a constant, say $w = t \in \mathbb{R}$ a.e. By the normalization of $w_k$ and strong $L^{p^*}$-convergence we deduce $t \neq 0$. On the other hand $|\{|w_k - t| > \delta\}| \to 0$ as $k \to \infty$. For $\delta < |t|$ this contradicts $|\{w_k = 0\}| \geq \theta$. □

**Step 7.8** *Exclusion of vanishing.*

*Proof.* Suppose on the contrary that

$$\sup_{R>0} \limsup_{\varepsilon\to 0} \sup_{x\in\mathbb{R}^n} \int_{B_x^R} F(\widetilde{w}_\varepsilon) = 0.$$

By the local generalized Sobolev inequality and the fact that $u_\varepsilon$ is extremal this implies

$$\begin{aligned} F^* &= \inf_{R>0} \liminf_{\varepsilon\to 0} \inf_{x\in\mathbb{R}^n} \int_{\mathbb{R}^n\setminus B_x^R} F(\widetilde{w}_\varepsilon) \\ &\leq \inf_{r>0} \liminf_{\varepsilon\to 0} \inf_{x\in\mathbb{R}^n} F^* \left(1 - \int_{B_x^r} |\nabla \widetilde{w}_\varepsilon|^p\right)^{p^*/p} \end{aligned}$$

or

$$\sup_{r>0} \limsup_{\varepsilon\to 0} \sup_{x\in\mathbb{R}^n} \int_{B_x^r} |\nabla \widetilde{w}_\varepsilon|^p = 0. \tag{7.2}$$

Another application of the local generalized Sobolev inequality yields vanishing with respect to a modified volume integrand, namely

$$\limsup_{\varepsilon\to 0} \sup_{x\in\mathbb{R}^n} |\{|\widetilde{w}_\varepsilon| \geq \delta\} \cap B_x^r| = 0 \text{ for every } \delta > 0,\ r > 0. \tag{7.3}$$

By hypothesis there exists a $t_0 > 0$ such that

$$\sup_{|t|\le t_0} \frac{F(t)}{|t|^{p^*}} \quad < \quad \frac{F^*}{S^*}. \tag{7.4}$$

Define

$$F_1(t) \quad := \quad \begin{cases} F(t) & (|t| < t_0), \\ 0 & (|t| \ge t_0), \end{cases} \qquad F_2 \quad := \quad F - F_1.$$

Then

$$F_2(t) \quad \le \quad F_3(t) \quad := \quad c\left(|t| - \frac{t_0}{2}\right)_+^{p^*}.$$

We claim that

$$\lim_{\varepsilon\to 0} \int_{\mathbb{R}^n} F_2(\widetilde{w}_\varepsilon) \quad = \quad 0. \tag{7.5}$$

To this end consider a cube $Q_z := z + [0,1)^n$ with $z \in \mathbf{Z}^n$ and observe that (7.3) implies

$$\lim_{\varepsilon\to 0} \sup_{z\in\mathbf{Z}^n} \left| \left\{ |\widetilde{w}_\varepsilon| > \frac{t_0}{2} \right\} \cap Q_z \right| \quad = \quad 0$$

i.e. that $F_3(\widetilde{w}_\varepsilon)$ vanishes on most of $Q_z$. Application of Lemma 7.7 yields

$$\int_{Q_z} F_3(\widetilde{w}_\varepsilon) \quad \le \quad c\left(\int_{Q_z} |\nabla\widetilde{w}_\varepsilon|^p\right)^{p^*/p}$$

for $\varepsilon$ small with a constant $c$ independent of $z$. After summation over $z \in \mathbf{Z}^n$ we find

$$\begin{aligned} \int_{\mathbb{R}^n} F_2(\widetilde{w}_\varepsilon) \quad \le \quad \int_{\mathbb{R}^n} F_3(\widetilde{w}_\varepsilon) \quad &\le \quad c \sum_{z\in\mathbf{Z}^n} \left(\int_{Q_z} |\nabla\widetilde{w}_\varepsilon|^p\right)^{p^*/p} \\ &\le \quad c \sup_{z\in\mathbf{Z}^n} \left(\int_{Q_z} |\nabla\widetilde{w}_\varepsilon|^p\right)^{\frac{p}{n-p}} \int_{\mathbb{R}^n} |\nabla\widetilde{w}_\varepsilon|^p . \end{aligned}$$

Now (7.5) follows from (7.2). In view of (7.4) we obtain the contradiction

$$F^* \quad = \quad \limsup_{\varepsilon\to 0} \int_{\mathbb{R}^n} F(\widetilde{w}_\varepsilon) \quad = \quad \limsup_{\varepsilon\to 0} \int_{\mathbb{R}^n} F_1(\widetilde{w}_\varepsilon) \quad < \quad F^*.$$

□

**Step 7.9** *Exclusion of concentration and convergence of* $(w_\varepsilon)$ *in* $D^{1,p}(\mathbb{R}^n)$ *to an entire extremal.*

*Proof.* By Lemma 7.5 and the Steps 7.6 and 7.8 there is a sequence of points $(x_\varepsilon)$ such that the measures $F(w_\varepsilon)$ are compact in the sense of the compactness-splitting-vanishing alternative. A subsequence of $(w_\varepsilon)$ satisfies the hypotheses of the generalized concentration compactness alternative (Theorem 4.11) with $\nu(\overline{\mathbb{R}^n}) = F^*$. Compactness in the sense of the compactness-splitting-vanishing alternative excludes concentration at infinity. Concentration at a finite point is excluded by the growth condition at infinity. Indeed, if $(w_\varepsilon)$ concentrates at a finite point then

$$F^* \quad = \quad \lim_{\varepsilon \to 0} \int_{\{|w_\varepsilon| \geq t\}} F(w_\varepsilon) \quad \leq \quad (F_\infty^+ + o(1))S^* \quad < \quad F^*$$

as $t \to \infty$. Thus alternative B of the generalized concentration compactness alternative (Theorem 4.11) occurs. □

**Step 7.10** *For a subsequence we have* $x_\varepsilon \to x_0$ *with* $x_0$ *as in the concentration theorem (Theorem 7.3).*

*Proof.* Let $v_\varepsilon = u_\varepsilon/\varepsilon$. By Theorem 7.3 we have $|\nabla v_\varepsilon|^p \overset{*}{\rightharpoonup} \delta_{x_0}$ in $\mathcal{M}(\overline{\Omega})$ for a subsequence. On the other hand

$$0 \quad < \quad \int_{B_0^R} |\nabla w|^p \quad = \quad \lim_{\varepsilon \to 0} \int_{B_0^R} |\nabla w_\varepsilon|^p \quad = \quad \lim_{\varepsilon \to 0} \int_{B_{x_\varepsilon}^{R_\varepsilon}} |\nabla v_\varepsilon|^p$$

for $R$ large enough and $R_\varepsilon := \varepsilon^{\frac{p}{n-p}} R$. This inequality fails if $x_\varepsilon$ does not tend to $x_0$. □

The proof of Theorem 7.4 is complete. □

As an immediate consequence of the theorem on the limit shape of low energy extremals we obtain existence of entire extremals.

**Corollary 7.11 (Existence of entire extremals)** *If* (1.3) *and* (7.1) *hold then* $F^*$ *admits an extremal function.*

Theorem 7.4 applies to the volume integrand and the plasma integrand.

**Example 7.12** As an example we consider the volume integrand on the unit ball. In Section 3.3 we found $w(x) = K(|x|)$ outside a ball of radius $r$ with $K(r) = 1$. The rescaled low energy extremal $w_\varepsilon$ is the p-capacity potential of a ball $B_0^{r_\varepsilon}$ with respect to $B_0^{\varepsilon^{\frac{p}{n-p}}}$ with $r_\varepsilon$ such that

$$\|\nabla w_\varepsilon\|_p \quad = \quad (K(r_\varepsilon) - K(\varepsilon^{\frac{p}{n-p}}))^{1-p} \quad = \quad 1$$

leading to $r_\varepsilon \to r$ as $\varepsilon \to 0$ and

$$w_\varepsilon(x) \quad = \quad K(|x|) - K(\varepsilon^{\frac{p}{n-p}}) \quad = \quad w(x) + O(\varepsilon^{\frac{p}{p-1}})$$

in $B_0^{\varepsilon^{\frac{p}{n-p}}} \setminus B_0^{r_\varepsilon}$.

Theorem 7.4 also determines the limit shape of the *cores* $A_\varepsilon := \{u_\varepsilon \geq t\}$. It says that the cores are asymptotically spherical. More precisely we have the following general result. We adopt the notation of the theorem and we assume $w \geq 0$. The symmetric difference of two sets is denoted $A \triangle B$.

**Corollary 7.13 (Limit shape of cores)** *There is a level $t_0 > 0$ such that for $0 < t < t_0$ we have*

$$\lim_{\varepsilon \to 0} \frac{\left|\{u_\varepsilon > t\} \triangle B^{r_\varepsilon}_{x_\varepsilon}\right|}{\left|B^{r_\varepsilon}_0\right|} = 0$$

*where $r_\varepsilon = \varepsilon^{\frac{p}{n-p}} r$ and $w(r) = t$. If $F$ is non-decreasing on $\mathbb{R}^+$ we can choose $t_0 = \max w$.*

*Proof.* By the decay properties of entire extremals (Theorem 6.5) there is a level $t_0 > 0$ below which $w$ is strictly decreasing. The assertion follows from $w_\varepsilon \to w$ in $L^{p^*}(\Omega)$; the last one from Lemma 6.4. □

Theorem 7.4 does not apply to maximizing sequences for the critical power integrand. In fact they can concentrate arbitrarily fast (Lions [91, Theorem I.1]). In Theorem 7.4 this behaviour is ruled out by the growth condition at infinity which permits us to prescribe the rate of concentration. Nevertheless the conditions on $F_0^+$ and $F_\infty^+$ allow for critical growth at 0 and infinity as shown by the following example. Let

$$F(t) := G(t)\,|t|^{p^*}, \quad G(t) = \begin{cases} 1 & (t \leq 1 \text{ or } t \geq 4), \\ 2 & (2 \leq t \leq 3) \end{cases}$$

and $1 \leq G(t) \leq 2$ otherwise. Then $F_0 = 1$, $F_\infty = 1$, and the conditions (7.1) on $F_0^+$ and $F_\infty^+$ is satisfied because $\int_{\mathbb{R}^n} F(u) > S^*$ for every extremal for $S^*$ with $\max u \geq 2$.

## 7.3 Exploiting the Euler Lagrange equation

In the presence of a smooth integrand further information on the asymptotic behaviour of low energy extremals can be derived from the Euler Lagrange equation

$$(7.6) \qquad \begin{aligned} -\operatorname{div}\left(|\nabla u_\varepsilon|^{p-2}\,\nabla u_\varepsilon\right) &= \lambda_\varepsilon f(u_\varepsilon) \text{ in } \Omega, \\ u_\varepsilon &= 0 \text{ on } \partial\Omega \end{aligned}$$

For the Lagrange multiplier we have the following obvious formula.

**Proposition 7.14 (Lagrange multiplier)** *If $F$ is differentiable,* (1.3) *and* (7.1) *holds, and $|f(t)| \leq c(1+|t|^{p^*})$ then the Lagrange multiplier in* (7.6) *is given by*

$$\lambda_\varepsilon = \varepsilon^{-\frac{p^2}{n-p}} \left( \frac{1}{p^* F^*} + o(1) \right)$$

*as $\varepsilon \to 0$.*

*Proof.* The Euler Lagrange equation for $u_\varepsilon$ is equivalent to

$$-\operatorname{div}\left( |\nabla w_\varepsilon|^{p-2} \nabla w_\varepsilon \right) = \varepsilon^{\frac{p^2}{n-p}} \lambda_\varepsilon f(w_\varepsilon)$$

for the rescaled functions $w_\varepsilon$. Choosing a subsequence we can assume

$$\lim_{\varepsilon \to 0} \varepsilon^{\frac{p^2}{n-p}} \lambda_\varepsilon = \overline{\lambda} \text{ in } \overline{\mathbb{R}^+}.$$

By Theorem 7.4 we have $w_\varepsilon \to w$ in $D^{1,p}(\mathbb{R}^n)$ and $L^{p^*}(\mathbb{R}^n)$. Therefore $f(w_\varepsilon) \to f(w)$ in $L^1_{\text{loc}}(\mathbb{R}^n)$ by hypothesis on $f$. On the other hand

$$\operatorname{div}\left( |\nabla w_\varepsilon|^{p-2} \nabla w_\varepsilon \right) \to \operatorname{div}\left( |\nabla w|^{p-2} \nabla w \right)$$

in the sense of distributions. This excludes $\overline{\lambda} = \infty$ because otherwise $w = 0$. Thus

$$-\operatorname{div}\left( |\nabla w|^{p-2} \nabla w \right) = \overline{\lambda} f(w)$$

in the distributional sense and $\overline{\lambda} = \lambda$ with $\lambda$ as in (6.2). By uniqueness of this limit the selection of a subsequence was unnecessary. □

In the semilinear case with subcritical nonlinearities $L^\infty$-bounds can be derived by Moser iteration [97]. A more general result of this type is due to Gidas and Spruck [65].

**Proposition 7.15 ($L^\infty$ bound for low energy extremals)** *If $|f(t)| \leq c\,|t|^{q-1}$ with $q < 2^*$ then $(w_\varepsilon)$ is bounded in $L^\infty(\Omega)$. Hence so is $(u_\varepsilon)$. Moreover $(w_\varepsilon)$ is bounded in $C^{1,\alpha}_{\text{loc}}(\mathbb{R}^n)$ for every $\alpha < 1$.*

*Proof.* By the Sobolev embedding $D^{2,q_k}(\mathbb{R}^n) \subset L^{\frac{nq_k}{n-2q_k}}(\mathbb{R}^n)$ and the Calderón-Zygmund inequality we have

$$\begin{aligned} \|w_\varepsilon\|_{L^{(q-1)q_{k+1}}(\Omega_\varepsilon)} &\leq a_k \|D^2 w_\varepsilon\|_{L^{q_k}(\Omega_\varepsilon)} \\ &\leq b_k \|f(w_\varepsilon)\|_{L^{q_k}(\Omega_\varepsilon)} \leq c_k \|w_\varepsilon\|^{q-1}_{L^{(q-1)q_k}(\Omega_\varepsilon)} \end{aligned}$$

with

$$q_{k+1} = \frac{1}{q-1} \frac{nq_k}{n-2q_k} > \left( \frac{n-2}{n+2} \frac{n}{n-2q_k} \right) q_k$$

because $q$ is subcritical. Each of the above embeddings is critical. Thus the constants do not depend on the domain. In our case they are independent of $\varepsilon$. The factor in brackets is $> 1$ for $q_k > \frac{2n}{n+2}$. By Sobolev's inequality the sequence $(w_\varepsilon)$ is bounded in $L^{2^*}(\mathbb{R}^n)$. Thus we can start with $q_0 := \frac{2^*}{q-1} > \frac{2n}{n+2} > 1$. After finitely many iterations we arrive at $2 - n/q_k > 1 + \alpha$, hence $H^{2,q_k}(\Omega_\varepsilon) \subset C^{1,\alpha}_{\text{loc}}(\mathbb{R}^n) \cap L^\infty(\mathbb{R}^n)$. □

**Proposition 7.16 (Compactness in $C^{2,\alpha}$)** *Assume the hypotheses of Proposition 7.15. If $f$ is Lipschitz then $w_\varepsilon \to w$ in $C^{2,\alpha}_{\text{loc}}(\mathbb{R}^n)$ for a subsequence.*

*Proof.* By the $L^\infty$-bound of Proposition 7.15 and Ascoli's theorem $w_\varepsilon \to w_0$ in $C^{1,\alpha}_{\text{loc}}(\mathbb{R}^n)$. By Theorem 7.4 on the limit shape of low energy extremals we know that the limit domain is all of $\mathbb{R}^n$. Moreover $w_\varepsilon \to w_0$ in $C^{2,\alpha}_{\text{loc}}(\mathbb{R}^n)$ by Schauder's estimate and Lipschitz continuity of $f$. Since $w_\varepsilon \to w$ in $D^{1,2}(\mathbb{R}^n)$ we conclude that $w_0 = w$. □

Next we show that under suitable hypotheses low energy extremals have a unique maximum point. In general this point is not identical with the point $x_\varepsilon$ of Theorem 7.4. But we will see that these points are close to each other. For simplicity we assume $u_\varepsilon \geq 0$. Let $y_\varepsilon$ be an arbitrary critical point of $u_\varepsilon$. If $f$ is Lipschitz and $f \geq 0$ on $\mathbb{R}^+$ then $u_\varepsilon(y_\varepsilon) = w_\varepsilon(0) \to \max w$. This follows from Lemma 6.4 for monotone integrands saying that every entire extremal $w$ is strictly decreasing outside its maximum set. Moreover $w_\varepsilon \to w$ in $C^1_{\text{loc}}(\mathbb{R}^n)$ by Proposition 7.16. Thus all critical points of $w_\varepsilon$ tend to the maximum set of $w$. In particular all critical points of $u_\varepsilon$ are within a distance of order $O(\varepsilon^{\frac{2}{n-2}})$ from each other. If in addition $f$ is non-decreasing then $f(\max w) > 0$ (otherwise $f(w) \equiv 0$) and the maximum set of $w$ is a singleton. Moreover $w = w^*$ up to translation by Lemma 6.1. In this case we can choose $x_\varepsilon$ to be any critical point of $u_\varepsilon$ and every other critical point $x'_\varepsilon$ satisfies $|x'_\varepsilon - x_\varepsilon| = o(\varepsilon^{\frac{2}{n-2}})$.

**Proposition 7.17 (Uniqueness of maximum point)** *If $f$ is Lipschitz and non-decreasing then $u_\varepsilon$ has at most one critical point for $\varepsilon$ sufficiently small. In particular the cores $\{u_\varepsilon > t\}$ with $t > 0$ are simply connected with respect to $\Omega$.*

*Proof.* Suppose on the contrary that for a subsequence $w_\varepsilon$ has two critical points $y_\varepsilon$ and $y'_\varepsilon$. By the above they both tend to the origin. After a suitable rigid motion depending on $\varepsilon$ we can assume that the both lie on the $y_1$ axis. By $C^{2,\alpha}$-convergence this leads to the contradiction

$$0 = \lim_{\varepsilon \to 0} \frac{\partial^2 w_\varepsilon}{\partial y_1^2}(0) = \frac{\partial^2 w}{\partial y_1^2}(0) = -\frac{1}{n} f(w(0)) < 0.$$

□

Some of the readers may be more familiar with singular perturbation problems of the form

$$\begin{aligned} \mu \operatorname{div}(|\nabla u|^{p-2} \nabla u) + f(u) &= 0 \text{ in } \Omega, \\ u &= 0 \text{ on } \partial\Omega \end{aligned} \tag{7.7}$$

involving a small viscosity parameter $\mu$. Solutions of this equation are usually obtained by unconstrained variation of the functional

$$E(u) := \frac{\mu}{p} \int_\Omega |\nabla u|^p - \int_\Omega F(u).$$

The principal solution is obtained as the first mountain pass level of this functional on $D^{1,p}(\Omega)$ or by minimization of $E$ over the *Nehari manifold*

$$M := \left\{ u \in D^{1,p}(\Omega) : \mu \int_\Omega |\nabla u|^p = \int_\Omega u f(u) \right\}$$

as follows by testing the Euler Lagrange equation with $u$. This approach requires that the functional is bounded below on the Nehari manifold which is the case if $f(t)/t^{p-1} \to \infty$ as $|t| \to \infty$. The translation between the constrained and the unconstrained formulation is provided by the following relations. An extremal $u_\varepsilon$ of the constrained variational problem (1.2) solves (7.7) with

$$\begin{aligned} \mu &= \lambda_\varepsilon^{-1} = \varepsilon^{\frac{p^2}{n-p}} (p^* F^* + o(1)), \\ E(u_\varepsilon) &= \frac{\mu}{p} \varepsilon^p - \int_\Omega F(u_\varepsilon) = \frac{p}{n-p} F^* \varepsilon^{p^*} (1 + o(1)) \end{aligned}$$

as follows from the Propositions 7.2 and 7.14.

# Chapter 8

# Robin Functions

The importance of Robin functions has been pointed out in the introduction. In this chapter we define Robin functions for several operators and derive some of their properties. Each Robin function is defined in terms of the corresponding Green's function.

## 8.1 P-Robin function

The *p-Green's function* is defined as the solution of the singular Dirichlet problem

$$\begin{aligned} -\operatorname{div}\left(|\nabla G_x|^{p-2}\,\nabla G_x\right) &= \delta_x \text{ in } \Omega, \\ G_x &= 0 \text{ on } \partial\Omega. \end{aligned}$$

Note that only in the semilinear case the Green's function can be used as a kernel function to solve the Dirichlet problem. The p-Green's function can be decomposed into a radially symmetric singular part and a regular part

$$G_x = K_x - H_x$$

where $K_x(y) = K(|y-x|)$ denotes the fundamental singularity of the p-Laplacian at the point $x$ (2.4). For smooth domains the p-Green's function exists and the regular part is Hölder continuous up to the singularity. This follows from a result of Kichenassamy and Véron [73]. The definition of the p-Robin function relies on this fact.

**Definition 8.1 (p-Robin function, p-harmonic radius)** *The regular part of the p-Green's function evaluated at the singularity*

$$\tau(x) := H_x(x)$$

*is called p-Robin function of* $\Omega$ *at* $x$. *The p-harmonic radius* $\rho(x)$ *is defined by the relation*

$$K(\rho(x)) = \tau(x).$$

*A minimum point of the p-Robin function or maximum point of the p-harmonic radius is called a p-harmonic center of* $\Omega$.

The p-harmonic radius has been introduced by Levitskii [81]. Near the singularity the p-Green's function can be expanded as

$$G_x(y) = K_x(y) - \tau(x) + o(1) \text{ as } y \to x. \tag{8.1}$$

For a ball $B_0^R$ the p-Green's function at the center is $G_0(y) = K(|y|) - K(R)$. Hence $\tau(0) = K(R)$ and

$$\rho(0) = R. \tag{8.2}$$

In the conformal case

$$\rho(x)^n = \exp(-\alpha_n \tau(x)) \tag{8.3}$$

with $\alpha_n$ as in (2.11). The p-harmonic radius is a positive continuous function on $\Omega$. The p-harmonic radius of non-smooth domains can be defined as the pointwise maximum of the p-harmonic radius of all smooth subdomains. We summarize its further properties.

**Lemma 8.2 (Basic properties of p-harmonic radius)** *The p-harmonic radius has the following properties.*

1. *Monotonicity: If* $x \in \Omega \subset \widetilde{\Omega}$ *then* $\rho(x) \leq \widetilde{\rho}(x)$. *In particular* $\rho(x) \geq d(x) := \operatorname{dist}(x, \partial\Omega)$.
2. *Scaling: Under a dilation* $\widetilde{x} = sx$ *the p-harmonic radius scales as a length* $\widetilde{\rho}(\widetilde{x}) = s\rho(x)$.
3. *Isoperimetric property: Among all domains of equal volume the maximum of the p-harmonic radius is maximal for balls, i.e.*

$$\max_{\Omega} \rho \leq \max_{\Omega^*} \rho = R_\Omega$$

*where* $R_\Omega$ *denotes the radius of the ball* $\Omega^*$.

*Proof.* Monotonicity follows from $G_x \leq \widetilde{G}_x$ which implies $\tau(x) \geq \widetilde{\tau}(x)$. In combination with (8.2) we obtain the lower bound in terms of the distance to the boundary. The scaling property follows from

$$\begin{aligned} \widetilde{G}_{\widetilde{x}}(\widetilde{y}) &= s^{\frac{p-n}{p-1}} G_x(y) \text{ and} \\ K(\widetilde{\rho}(\widetilde{x})) &= \widetilde{\tau}(\widetilde{x}) = s^{\frac{p-n}{p-1}} \tau(x) = K(s\rho(x)). \end{aligned}$$

The proof of the isoperimetric property is delayed to Corollary 9.4. □

We can prove the subsequent additional properties only for $p = 2$ and $p = n$. However, we expect that they hold for general $p$. In Chapter 9 we will give some evidence to this conjecture. In two dimensions a conformal equivalence $h$ is given by a univalent holomorphic map. In higher dimensions conformal mappings are *Möbius transformations*, i.e. translations, rotations, dilations, inversions $x \mapsto x/|x|^2$ and their compositions. For these maps $|h'|^{\frac{n-2}{2}}$ is harmonic where $|h'(x)|$ denotes the local amplification factor.

**Lemma 8.3 (Further properties of the p-harmonic radius)** *For $p = 2$ and $p = n$ the p-harmonic radius has the following properties.*

1. *Conformal equivariance: Under a conformal equivalence $h : \Omega \to \widetilde{\Omega}$, $x \to \widetilde{x}$ the p-harmonic radius transforms according to $\widetilde{\rho}(\widetilde{x}) = |h'(x)|\, \rho(x)$.*

2. *The p-harmonic radius of the ball $B_0^R$ is $\rho(x) = R - \frac{|x|^2}{R}$; that of the complement $(B_0^R)^c$ is $\rho(x) = \frac{|x|^2}{R} - R$; that of the half space $\{x_1 > 0\}$ is $\rho(x) = 2x_1$. If $\Omega$ is convex then $\rho \leq 2\,d$ where $d$ denotes the distance from the boundary.*

3. *Boundary behaviour: The p-harmonic radius vanishes at every boundary point satisfying the exterior ball condition. In particular a domain of finite volume satisfying the exterior ball condition has a least one p-harmonic center. At every boundary point satisfying the double sided ball condition we have*

$$\rho(x) \quad = \quad 2\,d(x) + o(d(x)) \quad \text{as} \quad d(x) \to 0.$$

*Proof.* For $p = 2$ the following transformation rules hold. See Gilbarg and Trudinger [66] for the first one.

$$\begin{aligned} \Delta \widetilde{u} &= |h'|^{-\frac{n+2}{2}} \Delta \left( |h'|^{\frac{n-2}{2}} u \right), \\ G_{\widetilde{x}}(\widetilde{y}) &= |h'(x)|^{\frac{2-n}{2}} |h'(y)|^{\frac{2-n}{2}} G_x(y), \\ \widetilde{\tau} &= |h'|^{2-n} \tau. \end{aligned}$$

Our candidate for $G_x(y)$ is $|h'(x)|^{\frac{n-2}{2}} |h'(y)|^{\frac{n-2}{2}} G_{\widetilde{x}}(\widetilde{y})$. By the above transformation rule for the Laplacian it is harmonic in $\Omega \setminus \{x\}$. Moreover it has the correct asymptotic behaviour at the singularity because

$$G_{\widetilde{x}}(\widetilde{y}) \quad = \quad K(|\widetilde{y} - \widetilde{x}|) - H_{\widetilde{x}}(\widetilde{y}) \quad = \quad |h'(x)|^{2-n} K(|y - x|) + O(1).$$

Subtraction of the singular part $K_x$ and evaluation at $x$ yields the transformation rule for the Robin function and hence for the harmonic radius.

The n-Green's function is conformally invariant and the fundamental singularity has a logarithmic structure. By conformality of $h$ we can expand the n-Green's function near the singularity as

$$G_{\widetilde{x}}(\widetilde{y}) \quad = \quad K(|\widetilde{y} - \widetilde{x}|) + K(|h'(x)|) - K(\widetilde{\rho}(\widetilde{x})) + o(1)$$

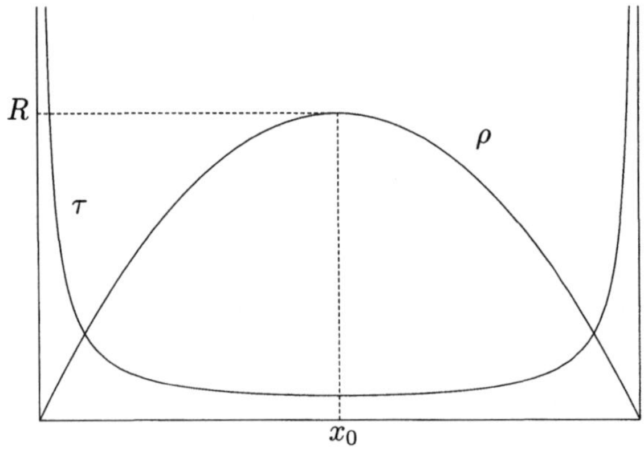


Figure 8.1: Robin function $\tau$, harmonic radius $\rho$ and harmonic center $x_0$ of 3 dimensional ball of radius $R$ plotted along a straight line through the center.

as $y \to x$. Comparison with the expansion (8.1) of $G_x$ near $x$ and $y \to x$ yields the transformation rule for the n-harmonic radius. The p-harmonic radius of the unit ball is obtained by choosing for $h$ a Möbius transformation of the unit ball to itself that maps the origin to a given point $x$. In complex coordinates on a two dimensional cross section containing the origin an the point $x$ it assumes the form

$$h(y) = \frac{y+x}{1+y\overline{x}} \tag{8.4}$$

with $|h'(0)| = 1 - |x|^2$. This is the p-harmonic radius of the unit ball. Composition with inversions and dilations yields the other formulas. The estimate for convex domains follows by comparison with half spaces. If $\Omega$ satisfies the exterior ball condition we can compare $\rho$ with the p-harmonic radius of $B^r_{x+r\nu} \subset \Omega^c := \mathbb{R}^n \setminus \overline{\Omega}$ which proves $\rho(x) = 0$. If it also satisfies the interior ball condition the p-harmonic radius of $\Omega$ can be pinched between that of $B^r_{x-r\nu}$ and that of $B^r_{x+r\nu}$ which implies $\nabla\rho(x) = -2\nu$. □

We expect that a strictly convex domain has a unique p-harmonic center. For the semilinear case in two dimensions this property is established in Lemma 8.6.

## 8.2 Robin function for the Laplacian

The regular part of the Green's function for the Laplacian is harmonic with the same boundary values as the singular part. This type of Robin function has been studied with C. Bandle [17]. The Green's function, the singular and the regular part are symmetric in $x$ and $y$. Figure 8.1 shows the Robin function and the harmonic radius of a ball in three dimensions. Further examples are calculated in [17].

**Lemma 8.4 (Properties of the harmonic radius)** *In addition to the properties of the p-harmonic radius the harmonic radius for the Laplacian has the following properties.*

1. *Regular boundary variation and gradient: The first variation of the Robin function and the harmonic radius subject to a smooth variation $\delta x$ of $\partial\Omega$ are given by*

$$\delta\tau(x) = -\int_{\partial\Omega} |\nabla G_x|^2 \, \delta x \cdot \nu, \qquad \delta\rho = \frac{\delta\tau}{K'(\rho)}.$$

2. *Boundary integral representation:*

$$\tau(x) = -\int_{\partial\Omega} K_x \partial_\nu G_x, \qquad \nabla\tau(x) = \int_{\partial\Omega} |\nabla G_x|^2 \, \nu$$

*Proof.* By Hadamard's variational formula for the Green's function (Garabedian [63]) we have

$$\delta G_x(y) = \int_{\partial\Omega} \nabla G_x \cdot \nabla G_y \, \delta x \cdot \nu.$$

Since the singular part $K_x$ does not vary this yields the first variation of the Robin function. By Green's identity

$$H_x(x) = -\int_{\partial\Omega} H_x \partial_\nu G_x$$

and $H_x = K_x$ on $\partial\Omega$ we obtain the formula for the Robin function. With $\delta x \cdot \nu = -\delta x_i$ (corresponding to a shift in the direction of the $i^{th}$ axis) we obtain the $i^{th}$ component of the gradient. □

A different boundary integral formula is due to Brézis and Peletier [28]. They found

$$\tau(x) = \frac{1}{n-2} \int_{\partial\Omega} |\nabla G_x|^2 \, (\cdot - x) \cdot \nu.$$

The proof uses regularization of the Green's function and Pohozaev's identity applied to the regularized problem.

## 8.3 Conformal radius and Liouville's equation

The harmonic radius of a simply connected planar domain is called *conformal radius.* The corresponding Robin functions solves a special elliptic boundary value problem discovered by Liouville [92]. A similar equation in higher dimensions is discussed in [17].

**Lemma 8.5 (Liouville's formula, Liouville's equation)** *In terms of a Riemann map $h : \Omega \to B_0^1$ the conformal radius is given by*

$$\rho = \frac{1 - |h|^2}{|h'|}.$$

*The corresponding Robin functions satisfies Liouville's equation*

$$\begin{aligned} \Delta\tau &= \frac{2}{\pi} e^{4\pi\tau} \quad \text{in } \Omega, \\ \tau &= +\infty \quad \text{on } \partial\Omega. \end{aligned}$$

*Proof.* If $h(x) = 0$ the Green's function at $x$ is

$$G_x = -\frac{1}{2\pi} \log |h|.$$

As $y \to x$ this implies

$$\begin{aligned} 2\pi\, G_x(y) &= -\log |h(y) - h(x)| \\ &= -\log |y - x| - \log |h'(x)| + O(|y - x|). \end{aligned}$$

We recognize that $\rho(x) = 1/|h'(x)|$. By composition with a Möbius transformation of the unit disk as in (8.4) we obtain Liouville's formula. Using complex coordinates $\partial_z = \frac{1}{2}(\partial_x - i\partial_y)$ and $\partial_{\bar{z}} = \frac{1}{2}(\partial_x + i\partial_y)$ Liouville's equation can be written as $\partial_{\bar{z}}\partial_z\tau = \frac{1}{2\pi}e^{4\pi\tau}$. This equation is easily verified using Liouville's formula. □

The Robin function is the *maximal solution* of Liouville's equation, i.e. every other solution is pointwise smaller than the Robin function. The maximal solution is unique and it is the only solution tending to infinity at the boundary as shown by Bandle, Essen, and Lazer [16, 80]. As a consequence of Liouville's equation we obtain that the conformal radius is a real analytic, positive, subharmonic function on $\Omega$. In particular the conformal radius has no relative minima in the interior of the domain. For a conformally flat metric in two dimensions $ds = \sigma\,|dx| = e^u\,|dx|$ the scalar curvature is given by

$$K = -\sigma^{-2}\Delta \log \sigma = -e^{-2\,u}\Delta u.$$

Thus Liouville's equation is Gauss's equation for the *hyperbolic metric* defined by the line element $ds := |dx|/\rho$. It has scalar curvature $K = -4$. By the boundary behaviour of the conformal radius (Lemma 8.3) the hyperbolic metric is complete on every domain satisfying the exterior ball condition. In the following lemma we use normal coordinates $(x, d) \in \partial\Omega \times \mathbb{R}^+$ representing $x - d\,\nu$ at a distance $d$ from the boundary.

**Lemma 8.6 (Properties of conformal radius)** *In addition to the properties of the harmonic radius the conformal radius satisfies:*

1. *Boundary behaviour: If $\partial\Omega$ is $C^{2,\alpha}$ then*

$$\rho(x - d\,\nu(x)) \;=\; 2\,d - H(x)d^2 + o(d^2),$$

*where $H$ denotes the mean curvature of the boundary with respect to the exterior normal.*

2. *Convexity (Kawohl [72, Example 1]): The Robin function of a (strictly) convex planar domain is (strictly) convex, the conformal radius is concave. In particular its level sets are convex. A convex domain which is not a strip has a unique conformal center.*

3. *The distance of a conformal center from the boundary is at least*

$$0.57088 \sup \rho$$

*(Minda and Overholt [95]).*

*Proof.* At the boundary the Laplacian assumes the form

$$\Delta\rho \;=\; \Delta_{\partial\Omega}\rho + \frac{\partial^2\rho}{\partial d^2} - H\frac{\partial\rho}{\partial d}.$$

We use this formula to compute the second normal derivative of the conformal radius. In terms of the conformal radius Liouville's equation assumes the form

$$\Delta\rho \;=\; \frac{|\nabla\rho|^2 - 4}{\rho}. \tag{8.5}$$

By Lemma 8.3 we already know that

$$\rho(x) \;=\; 2\,d(x) + o(d(x)) \quad \text{as} \quad d(x) \to 0.$$

Thus

$$\frac{|\nabla\rho|^2 - 4}{\rho} \;=\; \frac{(4 + 4\frac{\partial^2\rho}{\partial d^2}d + o(d)) - 4}{2d + o(d)} \;=\; 2\frac{\partial^2\rho}{\partial d^2} + o(1).$$

On the other hand

$$\Delta\rho \;=\; \frac{\partial^2\rho}{\partial d^2} - 2H.$$

Equating the two expressions for the Laplacian yields $\frac{\partial^2\rho}{\partial d^2} = -2H$. □

A more general convexity result is due to Bandle and Porru [20]. An example of a non-convex domain with multiple conformal centers is shown in Figure 8.2.

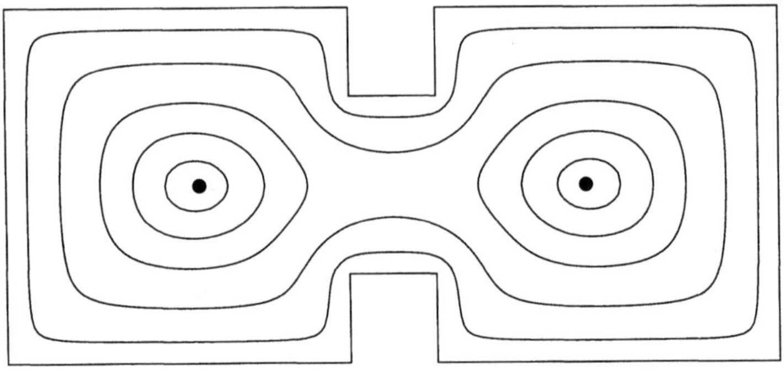

Figure 8.2: Level lines of Robin function on domain with two conformal centers.

## 8.4 Computation of Robin function and harmonic centers

In view of the applications of the Chapters 13, 14, and 15 it is of great importance to have an efficient method for the computation of the harmonic centers of a given domain, i.e. for the minimization of the Robin function. Evaluation of the Robin function according to its definition at a single point requires the solution of a Dirichlet problem. Typically a minimization process involves many evaluations of the function to be minimized. However, by exploiting the special properties of the Robin function we will be able to reduce the number of evaluations considerably. Moreover a suitable variant of the boundary element method allows us to evaluate the Robin function and its gradient simultaneously with little computational expense.

### 8.4.1 Boundary element method

To evaluate the Robin function $\tau(x) = H_x(x)$ at a point $x$ we need to solve

$$\begin{aligned} -\Delta H_x &= 0 \text{ in } \Omega, \\ H_x &= K_x \text{ on } \partial\Omega. \end{aligned}$$

However, we only need the value of $H_x$ at the point $x$. The boundary element method is designed to deal with this situation. It is based on the following boundary integral representation formula for a harmonic function $u$. The function $u^c$ is harmonic on the complement of $\Omega$.

(8.6)
$$\int_{\partial\Omega} K_y(\partial_\nu u - \partial_\nu u^c) - (u - u^c)\partial_\nu K_y$$

(8.7)
$$= \begin{cases} u(y) & (y \in \Omega), \\ \omega(y)u(y) + \omega^c(y)u^c(y) & (y \in \partial\Omega), \end{cases}$$

where

$$\partial_\nu K_y(z) = -\frac{(z-y)\cdot\nu(z)}{|S^{n-1}|\,|z-y|^n}.$$

At smooth boundary points we have $\omega(y) = \omega^c(y) = 1/2$. Otherwise

$$\omega(y) := \lim_{r\to 0}\frac{|\Omega\cap B_y^r|}{|B_y^r|}, \qquad \omega(y)^c := 1-\omega(y).$$

Formula (8.6) follows from Green's identity and the definition of $K_y$. The boundary element method only requires a triangulation of the boundary and provides high accuracy in the interior of the domain. It uses the second line of (8.6) to compute the complementary boundary values and the first line to evaluate the solution at an interior point. Depending on the choice of $u^c$ we obtain the following schemes.

**Single layer representation** With $u^c = u$ on $\partial\Omega$ equation (8.6) becomes

$$\int_{\partial\Omega} K_y\,E = u(y) \quad (y\in\overline{\Omega})$$

with $E = \partial_\nu u - \partial_\nu u^c$. Note that no computational effort is needed for the right hand side. For $y\in\partial\Omega$ the above is a Fredholm integral equation of the first kind for the unknown $E$ with a weakly singular, symmetric kernel. The corresponding bilinear form

$$\langle E, F\rangle := \int_{\partial\Omega} dy \int_{\partial\Omega} dz\,K_y(z)\,E(y)F(z)$$

is strictly positive definite since $\langle E,E\rangle = \int_\Omega |\nabla u|^2 + \int_{\Omega^c}|\nabla u^c|^2$. Thus the boundary equation can be solved by a standard Galerkin method leading to an approximate solution which is constant on each boundary element $\Gamma_i$. The coefficients of the resulting linear system are

$$A_{ij} = \int_{\Gamma_i} dy \int_{\Gamma_j} dz\,K_y(z).$$

In view of the singularity of the kernel an accurate quadrature formula is needed for the entries close to the diagonal. For a quasi-uniform triangulation of mesh-size $h$ the condition number is of order $O(h^{-1})$ (Johnson [69, Section 10.3.1]). For the computation of the Robin function we solve

$$\int_{\partial\Omega} K_y\,E_x = K_x(y) \quad (y\in\partial\Omega)$$

for $E_x$ and evaluate

$$\begin{aligned}\tau(x) &= \int_{\partial\Omega} K_x\,E_x,\\ \nabla\tau(x) &= -2\int_{\partial\Omega} \nabla K_x\,E_x,\end{aligned}$$

where we used $\nabla\tau(x) = 2\nabla H_x(x)$ and $\nabla_x K_x(z) = -\nabla_z K_x(z)$.

**Single layer representation with normal derivatives** With $u^c \equiv 0$ the representation formula (8.6) assumes the form

$$\int_{\partial\Omega} K_y\, \partial_\nu u \;=\; \begin{cases} u(y) + \int_{\partial\Omega} u\, \partial_\nu K_y & (y \in \Omega), \\ \omega(y)u(y) + \int_{\partial\Omega} u\, \partial_\nu K_y & (y \in \partial\Omega). \end{cases}$$

For the approximation of the Robin function we need to solve

$$\int_{\partial\Omega} K_y\, \partial_\nu H_x \;=\; \omega(y)K_x(y) + \int_{\partial\Omega} K_x\, \partial_\nu K_y, \quad y \in \partial\Omega$$

for the Neumann boundary values $\partial_\nu H_x$ and then evaluate

$$\tau(x) \;=\; \int_{\partial\Omega} K_x\, (\partial_\nu H_x - \partial_\nu K_x)\,.$$

Surprisingly we can also calculate its gradient from the data computed so far, namely

$$\nabla\tau(x) \;=\; \int_{\partial\Omega} (\partial_\nu H_x - \partial_\nu K_x)^2\, \nu.$$

This follows from Lemma 8.4.

**Double layer representation** With $\partial_\nu u^c = \partial_\nu u$ equation (8.6) becomes

$$\omega^c(y)\sigma(y) - \int_{\partial\Omega} \sigma\, \partial_\nu K_y \;=\; u(y) \quad (y \in \overline{\Omega})$$

with $\omega^c = 0$ in the interior of $\Omega$. On the boundary this is a Fredholm integral equation of the second kind for $\sigma = u - u^c$ with weakly singular, non-symmetric kernel. It has the advantage that the condition number of the corresponding linear system is bounded as $h \to 0$ and, remarkably enough, the integral equation can be solved to a specified accuracy with $O(N)$ operations, where $N$ denotes the number of boundary elements (Rokhlin [114]). In the case of the Robin function we solve

$$\omega^c(y)\sigma_x(y) - \int_{\partial\Omega} \sigma_x\, \partial_\nu K_y \;=\; K_x(y) \quad (y \in \partial\Omega)$$

for $\sigma_x$ and evaluate

$$\begin{aligned} \tau(x) &= -\int_{\partial\Omega} \sigma_x\, \partial_\nu K_x, \\ \nabla\tau(x) &= -2\int_{\partial\Omega} \sigma_x\, \nabla_x \partial_\nu K_x. \end{aligned}$$

**Capacity potentials** Also capacities can be easily computed by the boundary element method. Since

$$\operatorname*{cap}_{\Omega}(A) \;=\; -\int_{\partial A} \partial_\nu u_A \;=\; -\int_{\partial\Omega} \partial_\nu u_A$$

we only need the normal derivative of the capacity potential on one of the boundaries. The representation formula (8.6) reduces to

$$\int_{\partial(\Omega\setminus A)} K_y\, \partial_\nu u_A \;=\; \begin{cases} 0 & (y \in \partial\Omega), \\ 2\,\omega(y) & (y \in \partial A), \end{cases}$$

because

$$\int_{\partial A} \partial_\nu K_y \;=\; \begin{cases} 0 & (y \in \partial\Omega), \\ -\,\omega(y) & (y \in \partial A). \end{cases}$$

Solving this equation for $\partial_\nu u_A$ on $\partial A$ or $\partial\Omega$ thus permits us to compute $\operatorname{cap}_\Omega(A)$.

### 8.4.2 Computation of conformal radius

In addition to the methods for the harmonic radius the following methods can be used to compute the conformal radius.

**Conformal mapping** If a Riemann map $g : B_0^1 \to \Omega$ is known, Liouville's formula (Lemma 8.5) provides a parametrization of the conformal radius over the unit disk:

$$\rho(g(z)) \;=\; |g'(z)|\,(1-|z|^2).$$

For polygons with angles $\pi(1-\beta_k)$ a Riemann map is given by the Schwarz-Christoffel map

$$g(z) \;=\; g(0) + g'(0) \int_0^z \prod_{k=1}^{K} \left(1 - \frac{\xi}{z_k}\right)^{-\beta_k} d\xi$$

With this map we obtain

$$\rho(g(z)) \;=\; \rho(g(0)) \frac{1-|z|^2}{\prod_{k=1}^K |z - z_k|^{\beta_k}}.$$

A computer package of Trefethen [129, 128] can be used to determine the accessory parameters $z_k$ from the edges of the polygon. A new implementation of Trefethen's software in MATLAB has been made by Driscoll. In order to determine the conformal centers it is enough to find the maxima of the above ratio as $z$ varies over the unit disk and to evaluate $g$ at these points. For this purpose the value of the unknown factor $\rho(g(0))$ is irrelevant.

**Sinh formula** In terms of a complexification $\widetilde{G_x}$ of the Green's function the Riemann map can be written as $h = \exp(-2\pi\widetilde{G_x})$. Using

$$|h| = \exp(-2\pi G_x), \qquad |h'| = 2\pi \exp(-2\pi G_x)\,|\nabla G_x|$$

Liouville's formula assumes the form

$$\rho(y) = \frac{\sinh(2\pi G_x(y))}{\pi\,|\nabla G_x(y)|}.$$

The singularity at the point $x$ is removable. Using this formula we can compute the value of the conformal radius everywhere in $\Omega$ from the regular part $H_x$ for a single point $x$ by means of

$$\rho = 2\frac{\sinh(K_x - H_x)}{|\nabla K_x - \nabla H_x|}$$

with $H_x$ and its gradient computed by the boundary element method.

**Solution of Liouville's equation** The Robin function of a simply connected planar domain can be computed by solving Liouville's equation (Lemma 8.5). The only difficulty to be overcome are the infinite Dirichlet boundary values. The boundary behaviour of the conformal radius (Lemma 8.3) translates into

$$\tau(x) = -\frac{1}{2\pi}\log(2\,d(x)) + o(1) \quad \text{as} \quad d(x) \to 0. \tag{8.8}$$

With this information we can avoid the infinite boundary values. We replace $\Omega$ by the smaller domain

$$\Omega_h := \{x \in \Omega : d(x) > h\}.$$

Then we solve the approximate problem

$$\begin{aligned} \Delta\tau &= \frac{2}{\pi}e^{4\pi\tau} \quad \text{in } \Omega_h, \\ \tau &= -\frac{1}{2\pi}\log(2h) \quad \text{on } \partial\Omega_h. \end{aligned}$$

By monotonicity of the nonlinearity this problem is well posed as follows from the Leray-Schauder fixed point theorem. By (8.8) the approximation error at the boundary is $o(1)$ as $h \to 0$. By the maximum principle and monotonicity of the nonlinearity the same is true in all of $\Omega_h$. The approximate problem can be solved by Newton's method:

$$\begin{aligned} (-\Delta + 8e^{4\pi\tau_k})(\tau_{k+1} - \tau_k) &= \Delta\tau_k - \frac{2}{\pi}e^{4\pi\tau_k} \quad \text{in } \Omega_h, \\ \tau_{k+1} &= -\frac{1}{2\pi}\log(2h) \quad \text{on } \partial\Omega_h. \end{aligned}$$

By convexity of the nonlinearity an initial super solution leads to a pointwise decreasing sequence of super solutions. The Figures 8.2, 8.3, and 8.4 have been computed in this way.

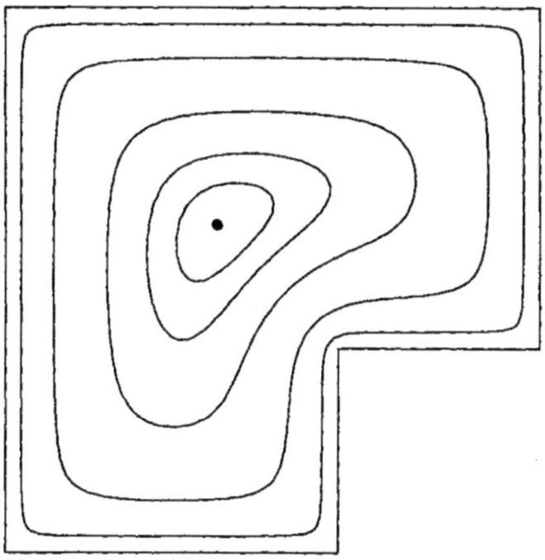

Figure 8.3: Level lines of Robin function and conformal center.

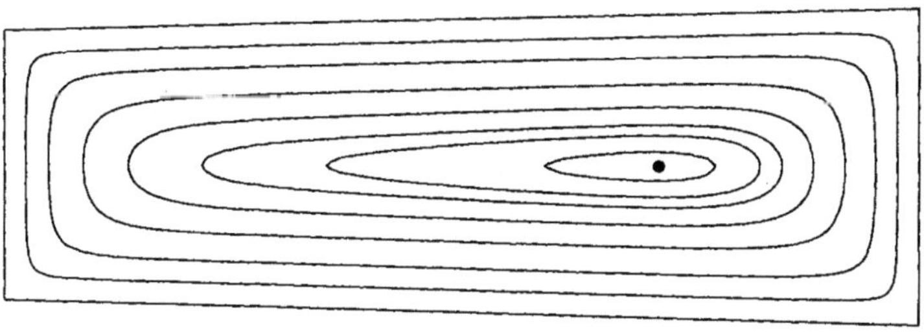

Figure 8.4: Domain with ill defined conformal center.

### 8.4.3 Computation of harmonic centers

To start the minimization procedure for the Robin function we proceed as follows. With an initial guess $x_0$ we make as if $\Omega$ were a ball of radius $r$ centered at $x_1$ with $r$ and $x_1$ to be determined. Thus

$$\begin{aligned}\rho(x_0) &= r - \frac{|x_1 - x_0|^2}{r},\\ \nabla\rho(x_0) &= \frac{2(x_1 - x_0)}{r}.\end{aligned}$$

This leads us to the simple iteration

$$x_1 = x_0 + \frac{2\,\rho(x_0)}{4 - |\nabla\rho(x_0)|^2}\nabla\rho(x_0).$$

Note that the boundary element method described in the previous section provides the values $\rho = K^{-1}(\tau)$ and $\nabla\rho = \nabla\tau/K'(\rho)$ at $x_0$ simultaneously. Once $\nabla\tau$ and $\tau$ are evaluated at several points we have enough information to proceed with a standard higher order minimization procedure like the quasi-Newton or the Fletcher-Reeves method.

Approximation of the maximal value of the conformal radius is a well conditioned problem. By Liouville's equation at least one of the second directional derivatives of the Robin function is large compared with the values of the Robin function itself. On the other hand the approximation of conformal centers may be ill posed as can be seen from Figure 8.4.

## 8.5 Other Robin functions

We first describe the construction of the Robin function and the corresponding harmonic radius for a general elliptic operator. Then we illustrate the general procedure by a number of examples. The harmonic radius for spherical and hyperbolic domains has been defined in [15].

Let $L$ be an arbitrary autonomous (independent of x) elliptic operator with a boundary operator $B$ for which the singular Dirichlet problem

$$\begin{aligned}L(G_x) &= \delta_x \text{ in } \Omega,\\ B(G_x) &= 0 \text{ on } \partial\Omega\end{aligned}$$

is well posed and for which the difference $G_x - \widetilde{G}_x$ of two solutions on two domains $\Omega$ and $\widetilde{\Omega}$ is continuous at $x$. This works at least for operators in divergence form $L(u) = -\mathrm{div}\,(a(|\nabla u|)\nabla u)$ with $a(t)t$ non-decreasing and $a(t)t \to \infty$ as $t \to \infty$. With $\widetilde{\Omega} = B_0^1$ we can define

$$\tau(x) := \lim_{y \to x}\left\{\widetilde{G}_x(y) - G_x(y)\right\}.$$

| $L$ | $-\Delta + k^2$ | $-\Delta + k^2$ | $-\Delta - k^2$ |
|---|---|---|---|
| $k^2$ | $\geq 0$ | $> 0$ | $< \lambda_1(\Omega)$ |
| $B$ | Dirichlet | Neumann | Dirichlet |
| $K(r)$ | $\frac{1}{8\pi s} e^{-ks}$ | $\frac{1}{8\pi s} e^{-ks}$ | $\frac{1}{8\pi s} \cos(ks)$ |
| $H_0(y)$ | $\frac{\sinh(k\|y\|)}{4\pi\|y\|(e^{2kr}-1)}$ | $\frac{-(1+k\,r)\sinh(k\,\|y\|)}{4\pi\|y\|(1+k\,r+e^{2\,k\,r}(k\,r-1))}$ | $\frac{\sin(k\|y\|)}{8\pi\|y\|\tan(kr)}$ |
| $H_0(0)$ | $\frac{k}{4\pi(e^{2kr}-1)}$ | $\frac{-k(1+k\,r)}{4\pi(1+k\,r+e^{2\,k\,r}(k\,r-1))}$ | $\frac{k}{8\pi\tan(kr)}$ |
| $\rho$ | $\frac{1}{k}\log\left(1+\frac{k}{4\pi\tau}\right)$ | not explicit | $\frac{1}{k}\arctan\left(\frac{k}{\tau}\right)$ |

Table 8.1: Helmholtz harmonic radius in 3 dimensions.

Then we define the harmonic radius of $(L, B)$ on $\Omega$ by the equation

$$\tau(x) = \tau_{B_0^{\rho(x)}}(0). \tag{8.9}$$

The Robin function depends on the choice of the fundamental singularity, i.e. on the choice of the reference domain $\widetilde{\Omega}$. However, the harmonic radius does not. For linear operators the above procedure involves the following steps.

1. Compute a radial fundamental singularity $K$.

2. Solve the auxiliary problem

$$\begin{aligned} LH_0 &= 0 \text{ in } B_0^r, \\ BH_0 &= BK_0 \text{ on } \partial B_0^r. \end{aligned}$$

3. Compute the harmonic radius as a function of the Robin function by solving $\tau = H_0(0)$ for $\rho = r$.

For operators of the form $L(u) = -\text{div}\,(a(|\nabla u|)\nabla u)$ the solutions of the auxiliary problem are constants and (8.9) reduces to $\tau(x) = K(\rho(x))$.

### 8.5.1 Helmholtz harmonic radius

The fundamental singularity of the Helmholtz operator $L = -\Delta + k^2$ in 3 dimensions is

$$K(r) = \frac{1}{8\pi r} e^{-kr}.$$

The solutions of the auxiliary problem

$$\begin{aligned}(-\Delta + k^2)H_0 &= 0 \text{ in } B_0^r,\\ H_0 &= K(r) \text{ on } \partial B_0^r\end{aligned}$$

are

$$H_0(y) = \frac{\sinh(k\,|y|)}{4\pi\,|y|\,(e^{2kr}-1)}$$

with $H_0(0) = \frac{k}{4\pi(e^{2kr}-1)}$. Solving this equation for $r$ gives the *Helmholtz harmonic radius* in terms of the Helmholtz Robin function, namely

$$\rho(x) = \frac{1}{k}\log\left(1 + \frac{k}{4\pi\tau(x)}\right).$$

As $k \to 0$ it tends to the harmonic radius for the Laplacian $\rho = \frac{1}{4\pi\tau}$. Schwarz symmetrization decreases the Helmholtz capacity. In particular the maximum of the Helmholtz harmonic radius among all domains of equal volume is maximal in the radial case. Table 8.5.1 shows further variants of the Helmholtz operator.

### 8.5.2 Biharmonic radius

The biharmonic Green's function corresponding to the clamped plate equation is the solution of

$$\begin{aligned}\Delta^2 G_x &= \delta_x \text{ in } \Omega,\\ G_x &= 0 \text{ on } \partial\Omega,\\ \partial_\nu G_x &= 0 \text{ on } \partial\Omega.\end{aligned}$$

In 2 dimensions a fundamental singularity is given by

$$K(r) = \frac{1}{8\pi} r^2 \log(r).$$

It is $C^1$ at the singularity with $K(0) = 0$. Thus the biharmonic Robin function is

$$\tau(x) = -G_x(x).$$

The solutions of the auxiliary problem

$$\begin{aligned}\Delta^2 H_0 &= 0 \text{ in } B_0^r,\\ H_0 &= K(r) \text{ on } \partial B_0^r,\\ \partial_\nu H_0 &= K'(r) \text{ on } \partial B_0^r\end{aligned}$$

are

$$\begin{aligned}H_0(y) &= \frac{1}{8\pi}\left(\log(r) + \frac{1}{2}\right)|y|^2 - \frac{1}{16\pi}|r|^2,\\ H_0(0) &= -\frac{1}{16\pi} r^2\end{aligned}$$

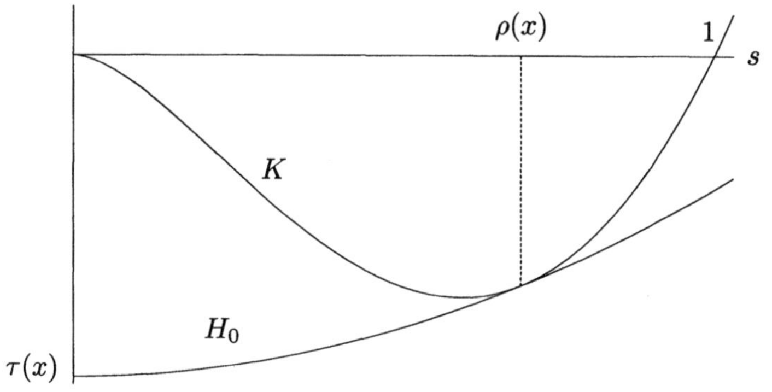


Figure 8.5: Definition of biharmonic radius.

(Figure 8.5). Thus the *biharmonic radius* is

$$\rho(x) = \sqrt{-16\pi\tau(x)}.$$

As an example we calculate the biharmonic radius of the unit disk. The biharmonic Green's function is

$$G_x(y) = \frac{1}{8\pi}\left(|y-x|^2\log\left(\frac{|y-x|}{|1-y\overline{x}|}\right)+\frac{1}{16\pi}\left(1-|x|^2\right)\left(1-|y|^2\right)\right)$$

(Garabedian [63]). Therefore the corresponding biharmonic Robin function and biharmonic radius are

$$\begin{aligned}\tau(x) &= -\frac{1}{16\pi}\left(1-|x|^2\right)^2, \\ \rho(x) &= 1-|x|^2.\end{aligned}$$

Surprisingly the biharmonic radius and the conformal radius of a disk are the same. For the numerical computation of the biharmonic centers we use the boundary element method. For this fourth order equation we also need the second and third derivative of $H_x$ at the boundary. These values are obtained from the fact that the Laplacian of a biharmonic function is harmonic.

# Chapter 9

# P-Capacity of Small Sets

The p-capacity of small sets is closely related to the p-harmonic radius. For the conformal radius this observation goes back to Teichmüller's habilitation thesis [124], perhaps even to Grötzsch. They found

$$\lim_{r\to 0}\left(\log(r)+\frac{2\pi}{\operatorname{cap}_\Omega(B_x^r)}\right) \;=\; \log(\rho(x)).$$

We prove a similar relation for the p-capacity (Theorem 9.2 below). It is based on elementary properties of the p-Green's function.

**Lemma 9.1 (Properties of p-Green's function)** *For fixed $x\in\Omega$ the p-Green's function satisfies*

$$\begin{aligned}\int_{\{G_x<t\}}|\nabla G_x|^p &= t,\\ \int_{\partial\{G_x>t\}}|\nabla G_x|^{p-1} &= 1,\\ \operatorname{pmod}_\Omega(\{G_x>t\}) &= t\end{aligned}$$

*for every $t>0$. The level sets of the p-Green's function on high levels are approximately small balls, i.e. as $t\to\infty$ we have $B_x^{r-}\subset\overline{\{G_x>t\}}\subset B_x^{r+}$ with $r$ defined by $t=K(r)-\tau(x)$ and $r_\pm=r\pm o(r^{\frac{n-1}{p-1}})$. In the semilinear case $r_\pm=r\pm O(r^n)$.*

*Proof.* Multiplication of $-\Delta G_x=\delta_x$ with the test function $\min(G_x,t)$ yields the first formula. Division by $t^p$ yields the third one; integration by parts the second one. On the level $t$ the expansion of the p-Green's function (8.1) assumes the form

$$t \;=\; K(r+\delta r)-\tau(x)+o(1) \;=\; t+K'(r)\delta r+o(1)$$

by definition of $r$. Hence $\delta r=o(1)/K'(r)$. In the semilinear case the regular part of the Green's function is smooth at $x$ and we can replace $o(1)$ by $O(r)$. □

Note that for subconformal $p$ the formula for $r_\pm$ is second order accurate while for $p = n$ it is only first order.

**Theorem 9.2 (P-Capacity of small balls)** *Assume $n \geq 2$ and fix $x \in \Omega$. Then the p-modulus and the p-capacity of a small ball centered at $x$ are asymptotically given by*

$$\begin{aligned}\operatorname{pmod}_\Omega(B_x^r) &= K(r) - \tau(x) + o(1),\\ \operatorname{pcap}_\Omega(B_x^r) &= \operatorname{pcap}_{\mathbb{R}^n}(B_x^r) + (p-1)\operatorname{pcap}_{\mathbb{R}^n}(B_x^r)^{\frac{p}{p-1}}(\tau(x) + o(1))\end{aligned}$$

*as $r \to 0$. In the radial case the error term in the first formula vanishes.*

*Proof.* For $r$ sufficiently small the gradient of $G_x$ does not vanish on $\partial B_x^r$. By the expansion of the p-Green's function (8.1) and Lemma 9.1 we can find a level $t = K(r) - \tau(x) + o(1)$ for which

$$\operatorname{pmod}_\Omega(B_x^r) \quad = \quad \operatorname{pmod}_\Omega(\{G_x > t\}) \quad = \quad t.$$

This yields the assertion. □

In the conformal case the formulas of the above theorem assume the form

$$\begin{aligned}\operatorname{nmod}_\Omega(B_x^r) &= K\left(\frac{r + o(r)}{\rho(x)}\right),\\ \operatorname{ncap}_\Omega(B_x^r) &= |S^{n-1}|\left(\log\frac{\rho(x)}{r}\right)^{1-n}\left(1 + o\left(\left(\log\frac{\rho(x)}{r}\right)^{-1}\right)\right)\end{aligned}$$

showing that the same formulas hold when the dimensionless quantity $r/\rho(x) \sim r/d(x)$ is small. Figure 9.1 compares the capacity of a small disk with respect to the unit disk to its approximation

$$\operatorname*{cap}_{B_0^1}(B_x^r) \quad \approx \quad -\frac{2\pi}{\log\left(\frac{r}{1-|x|^2}\right)}$$

given by Theorem 9.2. Together with the numerical methods of Section 8.4 the formula for the capacity of small balls applies to practical problems. For instance it can be used to approximate the Dirichlet eigenvalues of domains with small holes [50]. The increment of the Dirichlet eigenvalues, when a small set $A$ close to a point $x$ is removed, is given by

$$\lambda(\Omega \setminus A) \quad \approx \quad \lambda(\Omega) + \varphi^2(x)\operatorname*{cap}_\Omega(A)$$

where $(\lambda, \varphi)$ is a solution of the unperturbed eigenvalue problem

$$\begin{aligned}-\Delta\varphi &= \lambda\varphi \quad \text{in } \Omega,\\ \varphi &= 0 \quad \text{on } \partial\Omega,\\ \int_\Omega |\varphi|^2 &= 1.\end{aligned}$$

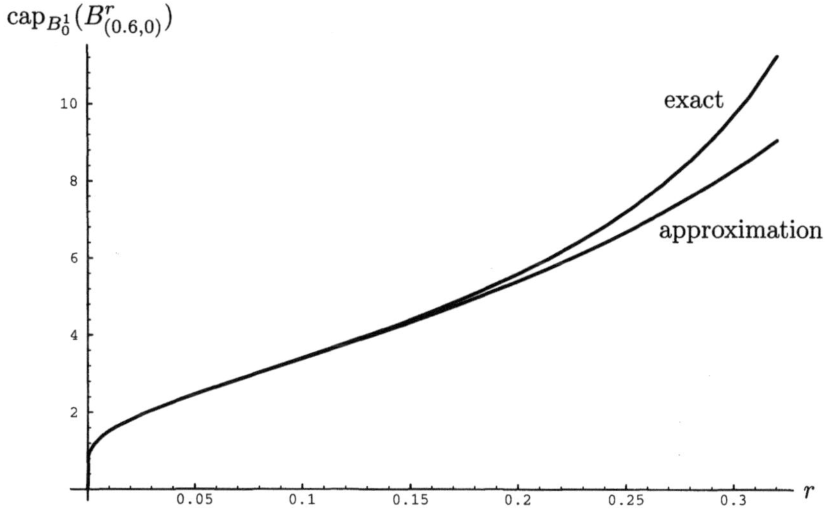


Figure 9.1: Accuracy of the formula for the capacity of small disks according to Theorem 9.2.

In the case of a small spherical hole we can approximate the capacity by the local term of Theorem 9.2. The resulting eigenvalue approximation formula turns out to be highly accurate [53].

The approximation formula for the p-capacity of small balls is equally useful to derive estimates for the p-harmonic radius that are sharper than that of Lemma 8.3.

**Corollary 9.3** *For every open set $A \subset \Omega$ and $x \in A$ we have*

$$\tau_A(x) \quad \geq \quad \tau_\Omega(x) + \text{pmod}_\Omega(A)$$

*with equality for the level sets of the p-Green's function $A = \{G_x > t\}$. In particular*

$$\text{pmod}_{\mathbb{R}^n}(A) \quad \leq \quad \inf \tau_A$$

*for $p < n$.*

*Proof.* By subadditivity of the p-modulus (Lemma 2.1) we have

$$\text{pmod}_\Omega(B_x^r) \quad \geq \quad \text{pmod}_A(B_x^r) + \text{pmod}_\Omega(A).$$

Together with the formula for the p-modulus of small balls we obtain

$$K(r) - \tau(x) + o(1) \quad \geq \quad K(r) - \tau_A(x) + o(1) + \text{pmod}_\Omega(A).$$

The assertion follows if we cancel the largest terms and let $r \to 0$. To prove equality for $A = \{G_x > t\}$ we use that the p-Green's function of the domain

$\{G_x > t\}$ at $x$ is $G_x - t$. The second inequality follows from the above if we set $\Omega = \mathbb{R}^n$ and observe that on $\mathbb{R}^n$ the subcritical p-Robin function vanishes identically. □

In the semilinear case in three dimensions a different proof for the second assertion can be found in Schiffer [116]. Next we prove the isoperimetric property of the p-harmonic radius. In the semilinear case it was first obtained by Pólya and Szegö for $n = 2$; in higher dimensions by Hersch [68].

**Corollary 9.4 (Isoperimetric property of p-harmonic radius)** *Among all domains of equal volume the maximum of the p-harmonic radius is maximal for balls, i.e.*

$$\sup \rho_\Omega \quad \leq \quad \sup \rho_{\Omega^*}.$$

*Proof.* For domains of infinite volume the assertion is trivial. Otherwise

$$\mathrm{pcap}_\Omega(B_x^r) \quad \geq \quad \mathrm{pcap}_{\Omega^*}(B_0^r)$$

by Schwarz symmetrization. By Theorem 9.2 we have

$$K(r) - K(\rho(x)) + o(1) \quad \leq \quad K(r) - K(\rho_{\Omega^*}(0))$$

where $\rho_{\Omega^*}(0) = R_\Omega$ is the radius of the ball $\Omega^*$. Thus $\rho(x) \leq R_\Omega$. □

Using the formula for the capacity of small balls we can give some evidence to our conjecture that the p-harmonic radius of balls is the same for all p and that Lemma 8.3 holds for general p. However, the argument is only heuristic. Under a conformal equivalence $h : \Omega \to \widetilde{\Omega}$ the p-Dirichlet integral transforms according to

$$\int_{\widetilde{\Omega}} |\nabla \widetilde{u}|^p \quad = \quad \int_\Omega |h'|^{n-p} \, |\nabla u|^p \,.$$

If we take for $u$ the p-capacity potential $u_r$ of a small ball $B_x^r$ we obtain

$$\mathrm{pcap}_{\widetilde{\Omega}}(h(B_x^r)) \quad \approx \quad |h'(x)|^{n-p} \, \mathrm{pcap}_\Omega(B_x^r)$$

because $|\nabla u_r|^p / \|\nabla u_r\|_p^p \overset{*}{\rightharpoonup} \delta_x$. By conformality the images $h(B_x^r)$ are asymptotic to a family of balls of radii $|h'(x)|\, r$. Using the formula for the capacity of small balls we obtain

$$\begin{aligned} \left(K(|h'(x)|\, r) - \widetilde{\tau}(\widetilde{x})\right)^{1-p} \quad &\approx \quad |h'(x)|^{n-p} \left(K(r) - \tau(x)\right)^{1-p}, \quad \text{or} \\ |h'(x)|^{\frac{p-n}{p-1}} \, K(r) - \widetilde{\tau}(\widetilde{x}) \quad &\approx \quad |h'(x)|^{\frac{p-n}{p-1}} \left(K(r) - \tau(x)\right). \end{aligned}$$

Cancellation of the leading terms and $\tau(x) = K(\rho(x))$ yields the transformation rule $\widetilde{\rho}(\widetilde{x}) = |h'(x)|\, \rho(x)$. The calculation is identical for $p = 2$ and for general $p$. The conclusion is correct for the semilinear case. Nevertheless, we are unable to estimate the error terms.

The next formula permits us to choose a simple reference domain for the computation of the p-capacity of a small set.

**Theorem 9.5 (Change of domain formula)** *If the sets $(A_\varepsilon)$ concentrate at a point $x \in \Omega \cap \widetilde{\Omega}$ in the sense $A_\varepsilon \subset B_x^{r_\varepsilon}$ with $r_\varepsilon \to 0$, then*

$$\operatorname{pmod}_\Omega(A_\varepsilon) = \operatorname{pmod}_{\widetilde{\Omega}}(A_\varepsilon) + \widetilde{\tau}(x) - \tau(x) + o(1)$$

*as $\varepsilon \to 0$.*

*Proof.* Denote by $u_\varepsilon$ the capacity potential of $A_\varepsilon$ with respect to $\Omega$. For some $y \in \Omega$ different from $x$ define

$$v_\varepsilon := \frac{G_x(y)}{u_\varepsilon(y)} u_\varepsilon.$$

Choose a compact connected set $K \subset \Omega \setminus \{x\}$ containing $y$ and fix $r > 0$ such that $K \cap \overline{B_x^r} = \emptyset$. For $\varepsilon$ sufficiently small $v_\varepsilon$ is a positive harmonic function on $\Omega \setminus \overline{B_x^r}$. By the *p-Harnack inequality* of Tolksdorf [127, Proposition 3.6.2] we have

$$\sup_K v_\varepsilon \leq c_K \inf_K v_\varepsilon \leq c_K G_x(y).$$

Thus the sequence $(v_\varepsilon)$ is bounded in $L^\infty(K)$. By a regularity theorem of Tolksdorf [127, Proposition 3.7] it is also bounded in $C^{1,\alpha}_{\text{loc}}(K)$ for some $\alpha > 0$. By Ascoli's compactness theorem and since $K$ was arbitrary we obtain $v_\varepsilon \to v$ in $C^{1,\beta}_{\text{loc}}(\Omega \setminus \{x\})$ for a subsequence with $\beta < \alpha$. By uniform convergence $v$ is a positive solution of

$$\begin{aligned} \int_\Omega |\nabla v|^{p-2} \nabla v \cdot \nabla \eta &= 0 \text{ for every } \eta \in C_c^\infty(\Omega \setminus \{x\}), \\ v &= 0 \text{ on } \partial\Omega \\ v(y) &= G_x(y). \end{aligned}$$

The only function with these properties is $v = G_x$. Now we fix a high level $t$ and define $r$ by $t = K(r) - \tau(x)$. By convergence of $v_\varepsilon$ we have $\{v_\varepsilon > t\} \approx \{G_x > t\}$ as $\varepsilon \to 0$ and $\{G_x > t\} \approx B_x^r$ as $t \to \infty$ by Lemma 9.1 on the p-Green's function. By additivity of the p-modulus (Lemma 2.1) in the case of p-capacity potentials we have

$$\begin{aligned} \operatorname{pmod}_\Omega(A_\varepsilon) &= \operatorname{pmod}_{\{v_\varepsilon > t\}}(A_\varepsilon) + \operatorname{pmod}_\Omega(\{v_\varepsilon > t\}) \\ &\approx \operatorname{pmod}_{B_x^r}(A_\varepsilon) + \operatorname{pmod}_\Omega(\{G_x > t\}) \\ &= \operatorname{pmod}_{B_x^r}(A_\varepsilon) + K(r) - \tau(x). \end{aligned}$$

We subtract the corresponding expression for $\widetilde{\Omega}$. Finally we first let $\varepsilon \to 0$ and then let $t \to \infty$. □

A simple choice for the reference domain is

$$\widetilde{\Omega} = \begin{cases} \mathbb{R}^n & (p < n), \\ B_x^1 & (p = n). \end{cases}$$

In both cases $\widetilde{\tau}(x) = 0$. For $p < n$ with $\widetilde{\Omega} = \mathbb{R}^n$ the change of domain formula can be written as

$$\text{(9.1)}\quad \operatorname{pcap}_\Omega(A_\varepsilon) = \operatorname{pcap}_{\mathbb{R}^n}(A_\varepsilon) + (p-1)\operatorname{pcap}_{\mathbb{R}^n}(A_\varepsilon)^{\frac{p}{p-1}}(\tau(x) + o(1)).$$

This generalizes the formula for the p-capacity of small balls (Theorem 9.2) to arbitrary sets of small diameter. The same should hold if we just assume that the p-capacity potentials of the sets $A_\varepsilon$ concentrate at $x$. A proof in the semilinear case is given in [51].

# Chapter 10

# P-Harmonic Transplantation

In this chapter we introduce the p-harmonic transplantation. This construction extends Hersch's harmonic transplantation [68] which in turn generalizes the conformal transplantation. Each of them is a powerful tool for the construction of comparison functions. They are complementary to symmetrization in the sense that they provide lower bounds when symmetrization gives upper bounds and vice versa. In contrast to symmetrization they take the geometry of the domain into account. This leads to sharper estimates.

**Definition 10.1 (P-Harmonic transplantation)** *Denote by $G_0$ the p-Green's function of $B_0^r$ with singularity at 0. To every positive radial function $U = \Phi \circ G_0 : B_0^r \to \mathbb{R}^+$ and $x \in \Omega$ we associate $u := \Phi \circ G_x : \Omega \to \mathbb{R}^+$. This transformation is called p-harmonic transplantation from $(B_0^r, 0)$ to $(\Omega, x)$.*

For the Laplacian on a simply connected planar domain this transformation reduces to the *conformal transplantation.* Given a Riemann map $g : B_0^1 \to \Omega$ with $g(0) = x$ it transforms a function $u$ defined on the unit disk into the function $u := U \circ g^{-1}$ defined on $\Omega$. By means of conformal transplantation Pólya and Szegö [104] proved the following inequality for the principal Dirichlet eigenvalue of a simply connected planar domain:

$$\lambda_1(\Omega) \leq \frac{\lambda_1(B_0^1)}{\sup \rho^2}$$

whereas symmetrization yields the opposite rough estimate $\lambda_1(\Omega) \geq \lambda_1(\Omega^*)$. The basic properties of the conformal transplantation they used are the conformal invariance of the Dirichlet integral and

$$\int_\Omega F(u) \geq \rho^2(x) \int_{B_0^1} F(U),$$

for every radial function $U$ and positive integrand $F$. This property is a consequence of the isoperimetric inequality

$$(10.1) \qquad \rho^2(x) = |g'(0)|^2 \leq \frac{1}{2\pi r} \int_{\partial B_0^r} |g'|^2$$

which follows from the fact that $|g'|^2$ is subharmonic. The bound for the principal Dirichlet eigenvalue is then obtained by choosing for $U$ the principal eigenfunction of the unit disk and $x$ a conformal center of $\Omega$. We give an extension of the inequality (10.1) to the p-Laplacian on a general domain.

**Lemma 10.2 (Lower bound for co-area of p-Green's function)** *Let $x \in \mathbb{R}^n$, $t > 0$ and $r > 0$ be given. Among all domains $\Omega$ with $\rho(x) = r$ the quantity*

$$\int_{\partial\{G_x > t\}} \frac{1}{|\nabla G_x|}$$

*is a minimum for $\Omega = B_x^r$.*

*Proof.* Let $\Omega^t := \{G_x > t\}$. By Hölder's inequality and Lemma 9.1 on the p-Green's function we can estimate

$$\begin{aligned}\int_{\partial\Omega^t} \frac{|\nabla G_x|^{\frac{p-1}{p}}}{|\nabla G_x|^{\frac{p-1}{p}}} &\leq \left(\int_{\partial\Omega^t} |\nabla G_x|^{p-1}\right)^{\frac{1}{p}} \left(\int_{\partial\Omega^t} \frac{1}{|\nabla G_x|}\right)^{\frac{p-1}{p}} \\ &= \left(\int_{\partial\Omega^t} \frac{1}{|\nabla G_x|}\right)^{\frac{p-1}{p}}.\end{aligned}$$

Thus the co-area of the p-Green's function is bounded below by

$$\int_{\partial\Omega^t} \frac{1}{|\nabla G_x|} \geq |\partial\Omega^t|^{\frac{p}{p-1}} \geq c_n \left|\Omega^t\right|^{\frac{p(n-1)}{n(p-1)}}$$

by the isoperimetric inequality relating surface and volume with optimal constant $c_n$. By the isoperimetric property of the p-harmonic radius (Corollary 9.4) we know that

$$\left|\Omega^t\right| \geq \left|B_0^{\rho_{\Omega^t}(x)}\right|.$$

The p-Green's function of $\Omega^t$ with singularity at $x$ is $G_x - t$. The corresponding value of the p-Robin function is

$$\tau_{\Omega^t}(x) = \tau(x) + t$$

which depends on $\Omega$ and $x$ only through the value $\tau(x) = K(\rho(x)) = K(r)$. Thus it remains unchanged if we replace $\Omega$ by $B_x^r$ leading to

$$\left|\Omega^t\right| \geq \left|(B_x^r)^t\right|.$$

In the radial case all inequalities of this proof are equalities and we can reverse the argumentation with the radial co-area at the end of the chain. □

**Theorem 10.3 (P-harmonic transplantation)** *The p-harmonic transplantation has the following properties.*

1. *It preserves the p-Dirichlet integral*

$$\int_\Omega |\nabla u|^p \;=\; \int_{B_0^r} |\nabla U|^p .$$

2. *If $r = \rho(x)$ then*

$$\int_\Omega F(u) \;\geq\; \int_{B_0^r} F(U).$$

*for every integrand $F : \mathbb{R} \to \mathbb{R}^+$. In the conformal case also $\int_\Omega F(u) \geq \rho(x)^n \int_{B_0^1} F(U)$.*

3. *For concentrating sequences the above inequality tends to an equality in the sense that if*

$$\begin{array}{lll} F(U_k) & \overset{*}{\rightharpoonup} & c\delta_0 \ \text{ in } \mathcal{M}(\overline{B_0^r}) \ \text{ then} \\ F(u_k) & \overset{*}{\rightharpoonup} & c\delta_x \ \text{ in } \mathcal{M}(\overline{\Omega}). \end{array}$$

*Proof.* Let $\Omega^t = \{G_x > t\}$. Using the co-area formula along the level sets of the p-Green's function the p-Dirichlet integral can be written as

$$\int_\Omega |\nabla u|^p \;=\; \int_0^\infty dt\; |\Phi'(t)|^p \int_{\partial\Omega^t} |\nabla G_x|^{p-1} \;=\; \int_0^\infty dt\; |\Phi'(t)|^p$$

which is independent of $\Omega$ and $x$. By Lemma 10.2 we have

$$\int_\Omega F(u) \;=\; \int_0^\infty dt\, F(\Phi(t)) \int_{\partial\Omega^t} \frac{1}{|\nabla G_x|} \;\geq\; \int_{B_0^r} F(U).$$

The main contribution to

$$\int_\Omega F(u_k) \;=\; \int_0^\infty dtF(\Phi_k(t)) \int_{\partial\Omega^t} \frac{1}{|\nabla G_x|}$$

stems from the large values of $t$. For those the isoperimetric inequality for the co-area of the p-Green's function asymptotically tends to an equality. □

P-harmonic transplantation can be used to construct approximate solutions of variational problems of the form (1.2) with small energy. In the semilinear case we propose the following procedure.

1. Find a harmonic center $z$ of the given domain and the corresponding value $R = \rho(z)$ of the harmonic radius by the method described in Section 8.4.3. Using one of the boundary element methods of Section 8.4.1 for this purpose also provides the boundary data needed to evaluate the Green's function $G_z$ at any point in $\Omega$.

2. Solve the radial variational problem on $B_0^R$. The corresponding Euler Lagrange equation is a one dimensional boundary value problem of the form

$$\begin{aligned} U'' + \frac{n-1}{r}U' + \lambda f(U) &= 0 \text{ in } (0,R), \\ U'(0) &= 0, \\ U(R) &= 0 \end{aligned}$$

   with the Lagrange multiplier given by Proposition 7.14.

3. Transplant the solution $U$ to the domain $\Omega$ using harmonic transplantation centered at the harmonic center $z$, i.e.

$$u(y) := U(K^{-1}(G_z(y) + K(R)).$$

   Note that the Green's function of $B_0^R$ at 0 is $G_0 = K_0 - K(R)$.

In Chapter 11 we will see that this procedure provides excellent approximate solutions of small Dirichlet energy.

# Chapter 11

# Identification of Concentration Points, Subconformal Case

In order to identify the possible concentration points of low energy limits we expand the supremum (1.2) with respect to $\varepsilon$ to second order. The complete strategy has been explained in the introduction.

To simplify the proof we assume that the integrand vanishes near zero. This class of integrands includes the examples volume and plasma integrand. More general integrands require decay estimates for the rescaled low energy extremals similar to that for entire extremals. See [51] for the semilinear case.

**Theorem 11.1 (Subconformal concentration points)** *Assume the general hypotheses* (1.3) *and the conditions* (7.1) *on* $F_0^+$ *and* $F_\infty^+$ *together with* $F \equiv 0$ *near zero (probably unnecessary). Then*

$$\lim_{\varepsilon \to 0} \tau(x_\varepsilon) = \inf_\Omega \tau$$

*with* $x_\varepsilon$ *given by Theorem* 7.4. *If* $\Omega$ *has a p-harmonic center* $z$ *then* $x_\varepsilon \to z$ *for a subsequence and*

$$\int_\Omega F(u_\varepsilon) = \varepsilon^{p^*} F^* \left(1 - \frac{n(p-1)}{n-p} w_\infty^p \inf_\Omega \tau \ \varepsilon^{\frac{p}{p-1}} + o\left(\varepsilon^{\frac{p}{p-1}}\right)\right)$$

*where*

$$w_\infty = \inf\left\{\lim_{r \to \infty} \frac{w(r)}{K(r)} : w \text{ is a positive radial extremal for } F^*\right\} > 0.$$

*Let* $U_\varepsilon$ *denote a radial extremal of the corresponding problem on a ball of radius* $\rho(z)$ *and define* $\widetilde{u}_\varepsilon$ *by p-harmonic transplantation to* $(\Omega, z)$. *Then*

$$\int_\Omega F(u_\varepsilon) = \int_\Omega F(\widetilde{u}_\varepsilon)\left(1 + o(\varepsilon^{\frac{p}{p-1}})\right).$$

The decay rate $w_\infty$ is realized by the entire extremal $w$ of Theorem 7.4 on the limit shape of low energy extremals. For the volume integrand of Section 3.3 we have $w_\infty = 1$ as follows from (3.8). Application of Theorem 11.1 yields

$$\begin{aligned}&\sup\{|A| \,:\, \mathrm{pcap}_\Omega(A) \le \varepsilon^p\}\\ =\ &\frac{\varepsilon^{p^*}}{(n(n-p))^{\frac{n}{n-p}}\,|B_0^1|^{\frac{p}{n-p}}}\left(1-\frac{n(p-1)}{n-p}\inf_\Omega \tau\ \varepsilon^{\frac{p}{p-1}} + o\left(\varepsilon^{\frac{p}{p-1}}\right)\right).\end{aligned}$$

Small balls centered at a p-harmonic center can serve as approximate solutions of this variational problem. They satisfy the same second order formula as the exact solutions. This fact is illustrated in Figure 14.5. For the semilinear plasma integrand in three dimensions (Section 3.4) with $F(t) = (t-1)_+^2$ we have $w_\infty = 2/3$ according to (3.10). Theorem 11.1 yields

$$\int_\Omega (u_\varepsilon - 1)_+^2 \;=\; \frac{\varepsilon^6}{108\,\pi^4}\left(1-\frac{4}{3}\inf_\Omega \tau\ \varepsilon^2 + o(\varepsilon^2)\right).$$

## 11.1 Lower bound

For the proof of Theorem 11.1 we can assume $u_\varepsilon \ge 0$. Let $w$ be an arbitrary positive radial extremal for $F^*$ and $w_\infty := \lim_{r\to\infty} w(r)/K(r)$. We can also assume that $w = w^*$ is strictly decreasing outside a ball $B_0^{r_0}$. By Lemma 6.2 this is no restriction. Note that

$$F_\Omega^\varepsilon \;=\; \int_\Omega F(u_\varepsilon) \;=\; \varepsilon^{p^*}\int_{\Omega_\varepsilon} F(w_\varepsilon)$$

with $\Omega_\varepsilon$ as in Theorem 7.4. Comparison functions will be constructed on $\Omega_\varepsilon$ rather than $\Omega$ itself.

**Step 11.2 (Lower bound)** *As $\varepsilon \to 0$ we have*

$$F_\Omega^\varepsilon \;\ge\; F_{B_0^{\sup\rho}}^\varepsilon \;\ge\; \varepsilon^{p^*}F^*\left(1-\frac{n(p-1)}{n-p}w_\infty^p\inf_\Omega \tau\ \varepsilon^{\frac{p}{p-1}} + o\left(\varepsilon^{\frac{p}{p-1}}\right)\right).$$

*Proof.* Let $z$ be a p-harmonic center of $\Omega$ (Definition 8.1). P-harmonic transplantation (Chapter 10) yields $F_\Omega^\varepsilon \ge F^\varepsilon_{B_0^{\rho(z)}}$. For the second inequality we set

$$r_\varepsilon \;:=\; \varepsilon^{-\frac{p}{n-1}}\rho(z), \quad R_\varepsilon \;:=\; \varepsilon^{-\frac{p}{n-p}}\rho(z).$$

We define the comparison functions $W_\varepsilon \in D^{1,p}(B_0^{R_\varepsilon})$ by $W_\varepsilon = w$ in $B_0^{r_\varepsilon}$ and $\Delta W_\varepsilon = 0$ in $B_0^{R_\varepsilon} \setminus B_0^{r_\varepsilon}$. Since $\int_{\mathbb{R}^n\setminus B_0^r} F(w) = O(r^{-\frac{n}{p-1}})$ (Theorem 6.5) we have

$$\begin{aligned}\int_{B_0^{R_\varepsilon}} F(W_\varepsilon) \;&\ge\; \int_{\mathbb{R}^n} F(w) - \int_{\mathbb{R}^n\setminus B_0^{r_\varepsilon}} F(w)\\ &=\; F^* - O(r_\varepsilon^{-\frac{n}{p-1}}) \;=\; F^* - o\left(\varepsilon^{\frac{p}{p-1}}\right).\end{aligned}$$

In view of the desired assertion errors of this order can be neglected. Using the formula for the p-capacity of small balls (2.7) we can estimate the p-Dirichlet integral from above by

$$\begin{aligned}
\|\nabla W_\varepsilon\|_p^p &= 1 - \int_{\mathbb{R}^n \setminus B_0^{r_\varepsilon}} |\nabla w|^p + \int_{B_0^{R_\varepsilon} \setminus B_0^{r_\varepsilon}} |\nabla W_\varepsilon|^p \\
&\leq 1 - w(r_\varepsilon)^p \left( \mathrm{pcap}_{\mathbb{R}^n}(B_0^{r_\varepsilon}) - \mathrm{pcap}_{B_0^{R_\varepsilon}}(B_0^{r_\varepsilon}) \right) \\
&= 1 - w(r_\varepsilon)^p \left( (K(r_\varepsilon))^{1-p} - (K(r_\varepsilon) - K(R_\varepsilon))^{1-p} \right) \\
&= 1 + (p-1) \left( \frac{w(r_\varepsilon)}{K(r_\varepsilon)} \right)^p K(R_\varepsilon)(1 + o(1)) \\
&= 1 + (p-1) w_\infty^p \tau(z) \varepsilon^{\frac{p}{p-1}} + o\left( \varepsilon^{\frac{p}{p-1}} \right)
\end{aligned}$$

by definition of $w_\infty$ and because $K(R_\varepsilon) = \varepsilon^{\frac{p}{p-1}} \tau(z)$ by definition of the p-harmonic radius (Definition 8.1). Moreover the ratio $K(R_\varepsilon)/K(r_\varepsilon)$ tends to zero. We scale $W_\varepsilon$ with $s \leq 1$ such that $\|\nabla W_\varepsilon^s\|_p = 1$. By the scaling properties (3.1) and the above estimates we obtain

$$\begin{aligned}
F_\varepsilon^*(B_0^{\rho(z)}) \;\geq\; F_\varepsilon^*(B_0^{s\rho(z)}) \;&\geq\; \int_{B_0^{sR_\varepsilon}} F(W_\varepsilon^s) \\
&= \|\nabla W_\varepsilon\|_p^{-\frac{n}{n-p}} \int_{B_0^{R_\varepsilon}} F(W_\varepsilon) \\
&\geq F^* \left( 1 - \frac{n(p-1)}{n-p} w_\infty^p \tau(z) \varepsilon^{\frac{p}{p-1}} + o\left( \varepsilon^{\frac{p}{p-1}} \right) \right)
\end{aligned}$$

as desired. □

## 11.2 Upper bound and identification of concentration points

Let $w$ be the entire extremal of Theorem 7.4 and $W_\infty := \lim_{r \to \infty} w(r)/K(r)$. Due to our simplifying hypothesis we can now set

$$r_\varepsilon := r_0, \quad R_\varepsilon := \varepsilon^{-\frac{p}{n-p}} \rho(x_\varepsilon), \quad A_\varepsilon := \{ w_\varepsilon > W_\infty K(r_\varepsilon) \}.$$

**Step 11.3 (Upper bound)** *If $F \equiv 0$ near zero then*

$$F_\Omega^\varepsilon \leq \varepsilon^{p^*} F^* \left( 1 - \frac{n(p-1)}{n-p} W_\infty^p \tau(x_\varepsilon) \varepsilon^{\frac{p}{p-1}} + o\left( \varepsilon^{\frac{p}{p-1}} \right) \right)$$

*as $\varepsilon \to 0$.*

*Proof.* We define the comparison functions $W_\varepsilon \in D^{1,p}(\mathbb{R}^n)$ by $W_\varepsilon = w_\varepsilon$ in $A_\varepsilon$ extended by a p-harmonic function to all of $\mathbb{R}^n$. By choice of $r_\varepsilon$ we have

$$\int_{\mathbb{R}^n} F(W_\varepsilon) \;\geq\; \int_{\Omega_\varepsilon} F(w_\varepsilon) - \int_{\Omega_\varepsilon \setminus A_\varepsilon} F(w_\varepsilon) \;=\; \varepsilon^{-p^*} F_\Omega^\varepsilon - o\left( \varepsilon^{\frac{p}{p-1}} \right).$$

For the p-Dirichlet integral we use the change of domain formula (9.1). In the present scaling it reads

$$\mathrm{pcap}_{\Omega_\varepsilon}(A_\varepsilon) - \mathrm{pcap}_{\mathbb{R}^n}(A_\varepsilon) = (p-1)\mathrm{pcap}_{\mathbb{R}^n}(A_\varepsilon)^{\frac{p}{p-1}}\varepsilon^{\frac{p}{p-1}}\tau(x_\varepsilon)(1+o(1)).$$

We can apply it because $r_\varepsilon/R_\varepsilon \to 0$. Using this formula and $\mathrm{pcap}_{\mathbb{R}^n}(A_\varepsilon) \geq \mathrm{pcap}_{\mathbb{R}^n}(B_0^{r_\varepsilon})(1-o(1))$ we can estimate the p-Dirichlet integral by

$$\begin{aligned}
\int_{\mathbb{R}^n} |\nabla W_\varepsilon|^p &= \int_{A_\varepsilon} |\nabla w_\varepsilon|^p + W_\infty^p K(r_\varepsilon)^p \mathrm{pcap}_{\mathbb{R}^n}(A_\varepsilon) \\
&\leq 1 - W_\infty^p K(r_\varepsilon)^p \left(\mathrm{pcap}_{\Omega_\varepsilon}(A_\varepsilon) - \mathrm{pcap}_{\mathbb{R}^n}(A_\varepsilon)\right) \\
&\leq 1 - (p-1)W_\infty^p K(r_\varepsilon)^p \mathrm{pcap}_{\mathbb{R}^n}(A_\varepsilon)^{\frac{p}{p-1}}\tau(x_\varepsilon)\varepsilon^{\frac{p}{p-1}}(1+o(1)) \\
&= 1 - (p-1)W_\infty^p \tau(x_\varepsilon)\varepsilon^{\frac{p}{p-1}} + o\left(\varepsilon^{\frac{p}{p-1}}\right)
\end{aligned}$$

because $\mathrm{pcap}_{\mathbb{R}^n}(B_0^{r_\varepsilon})^{\frac{p}{p-1}} = K(r_\varepsilon)^{-p}$ and $w(r_\varepsilon)/K(r_\varepsilon) \to W_\infty$. By the generalized Sobolev inequality (Lemma 3.2) we conclude

$$\begin{aligned}
\varepsilon^{-p^*} F_\Omega^\varepsilon &\leq \int_{\mathbb{R}^n} F(W_\varepsilon) + o\left(\varepsilon^{\frac{p}{p-1}}\right) \\
&\leq F^*\left(1 - (p-1)W_\infty^p \tau(x_\varepsilon)\varepsilon^{\frac{p}{p-1}} + o\left(\varepsilon^{\frac{p}{p-1}}\right)\right)^{\frac{n}{n-p}}
\end{aligned}$$

which is the assertion. □

For general integrands we choose $r_\varepsilon := \varepsilon^{-\frac{p}{n-1}}\rho(x_\varepsilon)$ as for the lower bound. By Theorem 7.4 we know that $w_\varepsilon \to w$ locally. But this is not enough for the present purpose. We need that $w_\varepsilon$ is close to $w$ on domains of size $r_\varepsilon$. This requires decay estimates for $w_\varepsilon$ showing that in the relevant range $w_\varepsilon$ and $w$ have the same decay properties. Together with the fact that they are close on compact sets (considered as initial condition) the decay estimates imply that they are close on domains that are small compared with $R_\varepsilon$. By Theorem 6.5 on the decay rate of entire extremals the above choice for $r_\varepsilon$ also guarantees that we can replace $w$ outside of $B_0^{r_\varepsilon}$ by $W_\infty K_0$ without changing the p-Dirichlet integral too much. For the semilinear case this program has been carried out with S. Müller [51].

**Step 11.4 (Identification of concentration point)** *As $\varepsilon \to 0$ we have $\tau(x_\varepsilon) \to \inf_\Omega \tau$ and $W_\infty = w_\infty$.*

*Proof.* By the Steps 11.2 and 11.3 we have

$$\begin{aligned}
&F^*\left(1 - \frac{n(p-1)}{n-p}W_\infty^p \tau(x_\varepsilon)\varepsilon^{\frac{p}{p-1}} + o\left(\varepsilon^{\frac{p}{p-1}}\right)\right) \\
\geq\ &F^*\left(1 - \frac{n(p-1)}{n-p}w_\infty^p \inf_\Omega \tau\ \varepsilon^{\frac{p}{p-1}} + o\left(\varepsilon^{\frac{p}{p-1}}\right)\right).
\end{aligned}$$

After cancellation of the leading term we obtain

$$W_\infty^p \tau(x_\varepsilon) \;\leq\; w_\infty^p \inf_\Omega \tau \;(1 + o(1))$$

for every $w_\infty$. This shows that $W_\infty$ must be minimal. Since $w_\infty > 0$ we conclude $\tau(x_\varepsilon) \to \inf_\Omega \tau$. □

# Chapter 12

# Conformal Low Energy Limits

For $p = n$ the critical Sobolev exponent becomes infinite. But $D^{1,n}(\Omega) \not\subset L^\infty(\Omega)$. Instead $D^{1,n}$ embeds into the Orlicz space given by the *Trudinger-Moser inequality*. The analysis of low energy limits in the conformal case differs in many respects from the subconformal case.

**Lemma 12.1 (Trudinger [130], Moser [98])** *If $\Omega \subset \mathbb{R}^n$ is a domain of finite volume then*

$$\sup\left\{\int_\Omega \exp\left(\alpha\,|u|^{\frac{n}{n-1}}\right) \;:\; u \in D^{1,n}(\Omega),\ \|\nabla u\|_n \le 1\right\} \quad < \quad \infty$$

*if and only if $\alpha \le \alpha_n$ with $\alpha_n$ as in* (2.11).

For $\alpha < \alpha_n$ the proof of the Trudinger-Moser inequality is simple (today). We follow Moser (see McLeod and Peletier [94] for a different proof including the result of Carleson and Chang). After symmetrization it suffices to consider radial functions. Consider $\Omega = B_0^1$. The change of variables

$$w(t) \quad := \quad n^{n-\frac{1}{n}}\left|S^{n-1}\right|^{\frac{1}{n}} u\left(\exp\left(-t/n\right)\right)$$

leads to

$$\int_{B_0^1} |\nabla u|^n \quad = \quad \int_0^\infty |w'|^n,$$

$$\int_{B_0^1} \exp\left(\alpha\,|u|^{\frac{n}{n-1}}\right) \quad = \quad \left|B_0^1\right| \int_0^\infty \exp\left(\frac{\alpha}{\alpha_n}\,|w|^{\frac{n}{n-1}} - t\right) dt.$$

For every normalized function $u$ Hölder's inequality yields

$$w(t) \quad = \quad \int_0^t w' \quad \le \quad t^{\frac{n-1}{n}}\left(\int_0^t |w'|^n\right)^{\frac{1}{n}} \quad \le \quad t^{\frac{n-1}{n}},$$

$$\int_0^\infty \exp\left(\frac{\alpha}{\alpha_n}\,|w|^{\frac{n}{n-1}} - t\right) dt \quad \le \quad \frac{\alpha_n}{\alpha_n - \alpha}$$

which proves the Trudinger-Moser inequality for $\alpha < \alpha_n$. For the critical exponent the proof is more delicate. We do not enter it here. Using suitable multiples of radial n-capacity potentials Moser also shows that $\alpha_n$ is the largest exponent for which the inequality holds. Using the above coordinates the functions $w_k(t) := k^{\frac{n-1}{n}} \min(t/k, 1)$ are normalized and

$$\begin{aligned} \int_0^\infty \exp\left(\frac{\alpha}{\alpha_n}|w_k|^{\frac{n}{n-1}} - t\right) dt &\geq \int_k^\infty \exp\left(\frac{\alpha}{\alpha_n}k - t\right) dt \\ &= \exp\left(k\left(\frac{\alpha}{\alpha_n} - 1\right)\right) \end{aligned}$$

which tends to infinity if $\alpha > \alpha_n$. We will use a concentration compactness alternative for this integrand to prove existence of extremal functions for the Trudinger-Moser inequality and concentration of low energy extremals in the conformal case.

## 12.1 Concentration limit

For general integrands we assume the general hypotheses (1.3) with $\alpha = \alpha_n$. In contrast to the generalized Sobolev constant and the local Sobolev constant the corresponding quantities in the conformal case depend on the domain.

**Definition 12.2 (Concentration limit)** *In the conformal case we denote*

$$\begin{aligned} \mathcal{B}_\Omega &:= \left\{u \in D^{1,n}(\Omega) : \|\nabla u\|_n \leq 1\right\}, \\ F_\Omega^* &:= \sup\left\{\int_\Omega F(u) : u \in \mathcal{B}_\Omega\right\}. \end{aligned}$$

*We define the* concentration limit *of $F$ at $x$ as*

$$F_\Omega^{\mathrm{loc}}(x) := \sup\left\{\int_\Omega F(u_\varepsilon) : u_\varepsilon \in \mathcal{B}_\Omega,\ (u_\varepsilon) \text{ concentrates at } x\right\}.$$

These quantities are finite by the growth condition and the Trudinger-Moser inequality. The concentration limit is closely related to the n-harmonic radius.

**Theorem 12.3 (Concentration limit)** *Let $\Omega$ be a bounded domain in $\mathbb{R}^n$ satisfying the exterior ball condition and assume* (1.3) *with $\alpha = \alpha_n$. Then*

$$F_\Omega^{\mathrm{loc}}(x) = F_{B_0^{\rho(x)}}^{\mathrm{loc}}(0) = \rho(x)^n F_{B_0^1}^{\mathrm{loc}}(0)$$

*for $x \in \Omega$ and $F_\Omega^{\mathrm{loc}}(x) = 0$ for $x \in \partial\Omega$.*

*Proof.* Scaling of the independent variable (3.1) yields the second equality. N-harmonic transplantation (Theorem 10.3) yields $F_\Omega^{\mathrm{loc}}(x) \geq F_{B_0^{\rho(x)}}^{\mathrm{loc}}(0)$. For the opposite inequality we choose a sequence $(u_\varepsilon)$ realizing $F_\Omega^{\mathrm{loc}}(x)$. We first consider

$x \in \Omega$. If $\Omega$ is a simply connected planar domain we can use a Riemann map $h : B_0^1 \to \Omega$ with $h(0) = x$ to define $U_\varepsilon := u_\varepsilon \circ h$. Then $(U_\varepsilon)$ concentrates at the origin and therefore

$$\begin{aligned}\lim_{\varepsilon\to 0}\int_\Omega F(u_\varepsilon) &= \lim_{\varepsilon\to 0}\int_{B_0^1} F(U_\varepsilon)\,|h'|^2 \\ &= |h'(0)|^2 \lim_{\varepsilon\to 0}\int_{B_0^1} F(U_\varepsilon) \;\le\; \rho(x)^2 F^{\mathrm{loc}}_{B_0^1}(0).\end{aligned}$$

For general domains we argue as follows. Note that

$$\begin{aligned}\int_{\{|u_\varepsilon|\le 1\}} F(u_\varepsilon) &= \int_{\{|u_\varepsilon|\le 1\}\cap B_x^{r_\varepsilon}} F(u_\varepsilon) + \int_{\{|u_\varepsilon|\le 1\}\setminus B_x^{r_\varepsilon}} F(u_\varepsilon) \\ &\le c r_\varepsilon^n + o(1)\end{aligned}$$

tends to 0 if we choose $r_\varepsilon \to 0$ large enough. Thus we can replace $u_\varepsilon$ below the level 1 by the n-capacity potential of $A_\varepsilon := \{u_\varepsilon > 1\}$ without changing the limit of the functional. The resulting function is denoted by $v_\varepsilon$. We apply the change of domain formula (Theorem 9.5) with $\widetilde{\Omega} = B_x^{\rho(x)}$ such that $\mathrm{nmod}_{\widetilde{\Omega}}(A_\varepsilon) = \mathrm{nmod}_\Omega(A_\varepsilon) + o(1)$. By the logarithmic structure of the fundamental singularity a change of order $o(1)$ of the radius of $\widetilde{\Omega}$ leads to a change of the same order of the n-modulus. Thus we can achieve that

$$\mathrm{nmod}_{B_x^{\rho(x)+o(1)}}(A_\varepsilon) \;\ge\; \mathrm{nmod}_\Omega(A_\varepsilon)$$

by increasing the radius of $\widetilde{\Omega}$ by $o(1)$. This leads to $\|\nabla v_\varepsilon\|_n \le \|\nabla u_\varepsilon\|_n \le 1$. By definition of the concentration limit we obtain

$$\begin{aligned}\lim_{\varepsilon\to 0}\int_\Omega F(u_\varepsilon) &= \lim_{\varepsilon\to 0}\int_{A_\varepsilon} F(u_\varepsilon) \\ &\le F^{\mathrm{loc}}_{B^{\rho(x)+o(1)}}(x) \;=\; (\rho(x)^n + o(1))\, F^{\mathrm{loc}}_{B_0^1}(0)\end{aligned}$$

as desired. Now consider $x \in \partial\Omega$. We use the fact that every conformal transformation preserves the n-Dirichlet integral. Since $\Omega$ satisfies the exterior ball condition we can find a ball $B_y^r \subset \Omega^c$ with $x \in \partial B_y^r$. Inversion with respect to its boundary transforms $\Omega$ into a subset of $B_y^r$ and the sequence $(u_\varepsilon)$ into a sequence $(\widetilde{u}_\varepsilon)$ that concentrates at $x$ because $x$ is a fixed point of the inversion. Therefore we can find a sequence $(h_\varepsilon)$ of Möbius transformations of $B_y^r$ into itself such that $\widetilde{u}_\varepsilon \circ h_\varepsilon$ concentrates at $y$ and $|h_\varepsilon'(0)| \to 0$ as $\varepsilon \to 0$. For instance we can choose $h_\varepsilon$ such that $h_\varepsilon(0)$ is the center of mass of $F(\widetilde{u}_\varepsilon)$. The amplification factor of the inversion is bounded above on the preimage of $\Omega$ in $B_y^r$ because $\Omega$ is bounded. Thus we can find a constant $c$ such that

$$\int_\Omega F(u_\varepsilon) \;\le\; c\int_{B_y^r} F(\widetilde{u}_\varepsilon \circ h_\varepsilon)\,|h'|^n \;=\; c\left(|h_\varepsilon'(0)|^n + o(1)\right) F^{\mathrm{loc}}_{B_y^r}(y)$$

which tends to zero as $\varepsilon \to 0$. □

The formula for the concentration limit allows us to deduce non-trivial properties of the concentration limit from those of the n-harmonic radius. In particular we obtain

$$\limsup_{\varepsilon\to 0} F_\Omega(u_\varepsilon) \;\leq\; \left(\max_\Omega \rho^n\right) F^{\mathrm{loc}}_{B_0^1}(0)$$

whenever the sequence $(u_\varepsilon)$ concentrates somewhere. Furthermore a maximizing sequence can only concentrate at a n-harmonic center.

## 12.2 Conformal concentration compactness alternative

In the conformal case we cannot use n-capacity potentials as cut-off functions to isolate the atoms of some measure from the rest of the measure as we did in the subconformal case (Corollary 3.4). Instead we use cut-off functions of log-log-log type for the same purpose. The results of this section have been obtained in collaboration with M. Struwe [57].

**Lemma 12.4** *There is a constant $c > 0$ such that*

$$ab \;\leq\; \exp\left(\alpha_n a^{\frac{1}{n-1}}\right) + c\,b\,(\log b)^{n-1} + c(a+b)$$

*for every $a, b > 0$.*

*Proof.* For $b > b_0$ the supremum in the definition of the convex conjugate function

$$\sup_{a>0}\left\{ab - \exp\left(\alpha_n a^{\frac{1}{n-1}}\right)\right\} \;=\; \sup_{A>0}\left\{\left(\frac{A}{\alpha_n}\right)^{n-1} b - \exp(A)\right\}$$

is attained at some $A > 0$ given by $(n-1)\frac{A^{n-2}}{\alpha_n^{n-1}} b = \exp(A)$. Thus

$$A \;=\; c + (n-2)\log(A) + \log(b) \;\leq\; \frac{A}{2} + \log(b)$$

for $A > A_0$. Hence $A \leq 2\log(b)$ for $a > a_0$ and

$$ab \;\leq\; \exp\left(\alpha_n a^{\frac{1}{n-1}}\right) + cb\,(\log b)^{n-1}$$

for $b \geq b_0$, $a \geq a_0$. Otherwise $ab \leq a_0 b + b_0 a$. □

**Lemma 12.5** *There are a cut-off functions $\eta_k \in D^{1,\infty}(\mathbb{R}^n)$ such that $\eta_k = 1$ in $B_x^{r_k}$ and $\eta_k = 0$ outside $B_x^{R_K}$ such that $r_k/R_k + R_k \to 0$ and*

$$\int_\Omega |u|^n\,|\nabla\eta_k|^n \;\leq\; \int_{B_x^{R_k}} \exp\left(\alpha_n\,|u|^{\frac{n}{n-1}}\right) + o(1)$$

*as $k \to \infty$ uniformly with respect to $u \in \mathcal{B}_\Omega$.*

*Proof.* Suppose $x = 0$. By Lemma 12.4 applied to $a = |u|^n$ and $b = |\nabla\eta_k|^n$ we have

$$\int_\Omega |u|^n |\nabla\eta_k|^n \le \int_{B_x^{R_k}} \exp\left(\alpha_n |u|^{\frac{n}{n-1}}\right) + c\int_{\{|\nabla\eta_k|>1\}} |\nabla\eta_k|^n (\log|\nabla\eta_k|)^{n-1}$$
$$+ c\int_{B_x^{R_k}} |u|^n + c\int_{B_x^{R_k}} |\nabla\eta_k|^n .$$

The third term can be absorbed into the first one up to an error of order $O(R_k^n)$. Similarly the fourth term can be absorbed into the second one. Now we choose

$$\eta(r) := \log\log|\log(r)|$$

defined for $r < \exp(-e)$. Define $r_k$ and $R_k$ by $\eta(R_k) = k$, $\eta(r_k) = k+1$ and note that $r_k/R_k + R_k \to 0$ as $k \to \infty$. Moreover

$$\int_0^{\exp(-e)} r^{n-1} |\eta'|^n (\log|\eta'|)^{n-1}\, dr \le \int_0^{\exp(-e)} \frac{(3\log\log|\log(r)|)^{n-1}}{r(|\log(r)|\log|\log(r)|)^n} dr$$
$$\le c\int_1^\infty \frac{(\log t)^{n-1}}{\exp((n-1)t)t^n} dt < \infty$$

after substituting $r = \exp(-\exp(t))$. Thus

$$\eta_k(x) := \max(0, \min(1, \eta(|x|) - k))$$

satisfies $\int_{\mathbb{R}^n} |\nabla\eta_k|^n (\log|\nabla\eta_k|)^{n-1} \to 0$ as $k \to \infty$. □

**Definition 12.6 (Compactness in the interior of $\mathcal{B}_\Omega$)** *We say that the functional $u \mapsto \int_\Omega F(u)$ is compact in the interior of $\mathcal{B}_\Omega$ if $\limsup_{\varepsilon\to 0} \|\nabla u_\varepsilon\|_n < 1$ and $u_\varepsilon \rightharpoonup u$ weakly in $D^{1,n}(\Omega)$ implies $F(u_\varepsilon) \to F(u)$ in $L^1(\Omega)$ for a subsequence.*

For the *Trudinger-Moser integrand* $F(t) = \exp\left(\alpha_n |t|^{\frac{n}{n-1}}\right) - 1$ this property follows from boundedness of the functional on $\mathcal{B}_\Omega$ and Vitali's theorem. For this integrand the following concentration compactness alternative is due to Lions [91, p. 196]. We present a new proof that applies to general integrands. It is different from that of [48] for the planar case.

**Theorem 12.7 (Conformal CCA)** *Let $\Omega$ be a domain in $\mathbb{R}^n$ of finite volume. Assume (1.3) with $\alpha = \alpha_n$ and that $u \mapsto \int_\Omega F(u)$ is compact in the interior of $\mathcal{B}_\Omega$. If a sequence $(u_\varepsilon)$ in $\mathcal{B}_\Omega$ satisfies $u_\varepsilon \rightharpoonup u$ weakly in $D^{1,n}(\Omega)$ and $|\nabla u_\varepsilon|^n \overset{*}{\rightharpoonup} \mu$ then one of the following assertions holds true.*

- A *Concentration: $\mu = \delta_{x_0}$ for some $x_0 \in \overline{\Omega}$, $u = 0$, and $F(u_\varepsilon) \overset{*}{\rightharpoonup} \gamma\delta_{x_0}$ for some $\gamma \in \mathbb{R}^+$.*

- B *Compactness: $F(u_\varepsilon) \to F(u)$ in $L^1(\Omega)$.*

*Proof.* If $\mu$ is not a Dirac mass then $\mu(B_x^R) < 1$ for $R \le R_0$ and every $x \in \overline{\Omega}$. Choosing cut-off functions $\eta_k$ as in Lemma 12.5 we show that $\eta_k u_\varepsilon$ lies in the interior of $\mathcal{B}_\Omega$. In fact by (3.3), convergence of $u_\varepsilon \to u$ in $L^n(\Omega)$, and Lemma 12.5 we have

$$\begin{aligned}
&\limsup_{\varepsilon\to 0} \int_\Omega |\nabla(\eta_k u_\varepsilon)|^n \\
\le\ & (1+\beta)^{n-1} \limsup_{\varepsilon\to 0} \int_{B_x^{R_k}} |\eta_k|^n |\nabla u_\varepsilon|^n + \left(1+\frac{1}{\beta}\right)^{n-1} \int_{B_x^{R_k}} |u|^n |\nabla \eta_k|^n \\
\le\ & (1+\beta)^{n-1} \mu(B_x^{R_k}) + \left(1+\frac{1}{\beta}\right)^{n-1} \left(\int_{B_x^{R_k}} \exp\left(\alpha_n |u|^{\frac{n}{n-1}}\right) + o(1)\right)
\end{aligned}$$

as $k \to \infty$. Choosing $\beta$ small and $k$ large enough the right hand side is $< 1$. By compactness in the interior of $\mathcal{B}_\Omega$ we have $F(\eta_k u_\varepsilon) \to F(\eta_k u)$ in $L^1(\Omega)$ for a subsequence. Hence $F(u_\varepsilon) \to F(u)$ in $L^1(B_x^{r_k} \cap \Omega)$. Covering $\overline{\Omega}$ by finitely many balls $B_x^{r_k}$ yields $F(u_\varepsilon) \to F(u)$ in $L^1(\Omega)$.

In the case of concentration the weak limit is $u = 0$ and $F(u_\varepsilon) \overset{*}{\rightharpoonup} \gamma\delta_{x_0}$. In fact, by Rellich's theorem $u_\varepsilon \to u$ in $L^n(\Omega)$ and

$$\int_\Omega |(1-\eta_k)u|^n \le c \limsup_{\varepsilon\to 0} \int_{B_x^{R_k}} |\nabla((1-\eta_k)u_\varepsilon)|^n = o(1)$$

as $k \to \infty$ by the same estimates as above and $(1-\eta_k)(x_0) = 0$. As $k \to \infty$ we obtain $u_\varepsilon \to 0$ in $L^n(\Omega)$. For a subsequence the limit $\gamma := \lim \int_\Omega F(u_\varepsilon)$ exists. By the same estimate as above and compactness of $u \mapsto \int_\Omega F(u)$ in the interior of $\mathcal{B}_\Omega$ we have $F(u_\varepsilon) - F(0) \to 0$ in $L^1(\Omega \setminus B_{x_0}^{R_k})$ for a subsequence. The assertion follows as $k \to \infty$. □

The following compactness criterion is an immediate consequence of the conformal concentration compactness alternative.

**Corollary 12.8 (Conformal compactness criterion)** *If $u \mapsto \int_\Omega F(u)$ is compact in the interior of $\mathcal{B}_\Omega$ and*

$$\sup F_\Omega^{\text{loc}} < F_\Omega^*$$

*then $F_\Omega^*$ is attained.*

*Proof.* From a maximizing sequence $(u_\varepsilon)$ for $F_\Omega^*$ we choose a subsequence such that $u_\varepsilon \rightharpoonup u \in \mathcal{B}_\Omega$ and $|\nabla u_\varepsilon|^n \overset{*}{\rightharpoonup} \mu$ in the sense of measures. Concentration in the sense of the conformal concentration compactness alternative is excluded by hypothesis. Therefore $F_\Omega(u_\varepsilon) \to F_\Omega(u)$ and $u$ realizes $F_\Omega^*$. □

For certain nonlinearities there are more direct methods to prove single point concentration. As an example we consider the conformal plasma integrand with a strong nonlinearity. For $q \ge n$ denote by $u_\varepsilon$ a solution of (1.2) with $F(t) = (t-1)_+^q$. We show that the cores $A_\varepsilon := \{u_\varepsilon > 1\}$ are connected. This proves concentration because in the conformal case the diameter of a connected set is a coercive functional in the sense that the diameter tends to 0 as the n-capacity tends to 0 (2.12). If $A_\varepsilon$ were disconnected we could decompose it into two disjoint, nonempty open sets $A_\varepsilon^1$ and $A_\varepsilon^2$. Testing the Euler Lagrange equation for $u_\varepsilon$ on $A_\varepsilon^k$ with $u_\varepsilon - 1$ we get

$$\delta_\varepsilon^k := \int_{A_\varepsilon^k} |\nabla u_\varepsilon|^n = \lambda_\varepsilon \int_{A_\varepsilon^k} F(u_\varepsilon) \tag{12.1}$$

with the same Lagrange multiplier on both components. Define

$$v_\varepsilon(x) := \begin{cases} 1 + \beta_\varepsilon(u_\varepsilon(x) - 1) & (x \in A_\varepsilon^1), \\ 1 & (x \in A_\varepsilon^2), \\ u_\varepsilon(x) & (u_\varepsilon(x) \leq 1), \end{cases}$$

with $\beta_\varepsilon^n := (\delta_\varepsilon^1 + \delta_\varepsilon^2)/\delta_\varepsilon^1$. This function has the same n-Dirichlet integral as $u$ and by (12.1) it has at least the same value of the functional:

$$\begin{aligned} \int_\Omega F(v_\varepsilon) &= \beta_\varepsilon^q \int_{A_\varepsilon^1} F(u_\varepsilon) \\ &= \beta_\varepsilon^{q-n} \frac{\delta_\varepsilon^1 + \delta_\varepsilon^2}{\delta_\varepsilon^1} \int_{A_\varepsilon^1} F(u_\varepsilon) = \beta_\varepsilon^{q-n} \int_\Omega F(u_\varepsilon) \geq \int_\Omega F(u_\varepsilon), \end{aligned}$$

since $\beta \geq 1$ and $q \geq n$. Therefore $v_\varepsilon$ is extremal and solves the Euler Lagrange equation. On the other hand, by the strong maximum principle it is not $C^1$ across $\partial A_\varepsilon^2$. Moreover $A_\varepsilon$ cannot have holes that are entirely contained in $\Omega$. Thus it is simply connected with respect to $\Omega$.

## 12.3 Extremal functions for the Trudinger-Moser inequality

As a first application of the concentration limit and the conformal concentration compactness alternative we address the question whether the Trudinger-Moser inequality (Lemma 12.1) admits an extremal function. Carleson and Chang [32] proved existence of extremal functions in the radial case. This result is not extendable to general domains by symmetrization. But we will see that the ratio of the functional and its maximal concentration limit satisfies an isoperimetric inequality. By means of this inequality the general case can be reduced to the result of Carleson and Chang. Thus we obtain existence of extremal functions on arbitrary domains of finite volume. Existence of solutions to the Euler Lagrange equation of the Trudinger-Moser functional has been shown by Adimurthi [4]. However, this is only a necessary condition for the existence of extremal functions. The difficulty in finding a maximizer for the Trudinger-Moser functional stems from its lack of compactness, i.e. its discontinuity with respect to weak convergence in $D^{1,n}(\Omega)$. In fact the Palais Smale condition fails along the sequence of rescaled n-capacity potentials described at the beginning of this chapter. Thus the Trudinger-Moser functional is not compact on the boundary of $\mathcal{B}_\Omega$. Nevertheless it is compact in the interior. For the Trudinger-Moser functional on a ball the conformal compactness criterion of Corollary 12.8 has been verified by Carleson and Chang. This has been achieved by computing the maximal concentration limit explicitly and construction of a comparison function above this critical level. In the radial case the maximum of the concentration limit is achieved at the origin.

**Lemma 12.9 (Carleson and Chang [32])** *For the Trudinger-Moser funtional on the unit ball one has*

$$\sup F^{\text{loc}}_{B_0^1} = |B_0^1| \exp\left(\sum_{k=1}^{n-1} \frac{1}{k}\right), \qquad F^*_{B_0^1} > |B_0^1| \exp\left(\sum_{k=1}^{n-1} \frac{1}{k}\right).$$

Thus for balls the Trudinger-Moser inequality admits extremal functions. Stability of the conformal compactness criterion of Corollary 12.8 under perturbations of the domain extends this result to domains which are close to a ball (Struwe [120]). In general, however, replacing a ball by another domain of the same volume decreases both sides in the conformal compactness criterion by a factor which is not necessarily close to 1. Thus the conformal compactness criterion might fail. Fortunately the ratio $F^*_\Omega / \sup F^{\text{loc}}_\Omega$ can only increase.

**Lemma 12.10** *Assume* (1.3) *with* $\alpha = \alpha_n$. *Then*

$$\frac{F^*_\Omega}{\sup F^{\text{loc}}_\Omega} \geq \frac{F^*_{B_0^1}}{\sup F^{\text{loc}}_{B_0^1}}$$

*for every domain* $\Omega$ *in* $\mathbb{R}^n$ *of finite volume.*

*Proof.* Using n-harmonic transplantation (Theorem 10.3) this is an immediate consequence of the formula for the concentration limit (Theorem 12.3). □

Thus $F^*_\Omega / \sup F^{\text{loc}}_\Omega$ is minimal for balls. In particular the radial case is the worst case with respect to the compactness criterion. Equality holds if and only if $\Omega$ is a ball. Lemma 12.10 makes the conformal compactness criterion much more applicable, because verifying the conformal compactness criterion in the radial case is a 1-dimensional problem. Combining the conformal concentration compacntess alternative with the above isoperimetric inequality and the result of Carleson and Chang we obtain existence of extremal functions for the Trudinger-Moser inequality.

**Corollary 12.11 (Extremals for the Trudinger-Moser inequality)** *On every domain* $\Omega \subset \mathbb{R}^n$ *of finite volume the Trudinger-Moser inequality admits an extremal function.*

For $n = 2$ this result has been obtained in [48], for $n \geq 3$ by K.C. Lin [88].

## 12.4 Concentration of conformal low energy extremals

For the conformal case we do not have a general result in the spirit of Theorem 11.1 to identify the concentration points of low energy limits. We expect the following to be true.

**Conjecture 12.12 (Conformal low energy limits)** *Let $\Omega$ be a bounded domain in $\mathbb{R}^n$ satisfying the exterior ball condition and assume the hypotheses* (1.3) *with $p = n$. For $\varepsilon$ small enough the variational problem* (1.2) *has a solution $u_\varepsilon$. Denote by $v_\varepsilon$ the solution of the corresponding radial problem on the unit ball $B_0^1$. Then $(u_\varepsilon)$ concentrates at a n-harmonic center of $\Omega$ and*

$$\int_\Omega F(u_\varepsilon) \;=\; \left(\max_\Omega \rho^n + o(1)\right) \int_{B_0^1} F(v_\varepsilon).$$

Existence of the extremal functions $u_\varepsilon$ and $v_\varepsilon$ follows from the growth condition, the Trudinger-Moser inequality, and upper semi continuity of the integrand. Using n-harmonic transplantation we obtain the inequality

$$\int_\Omega F(u_\varepsilon) \;\geq\; \max_\Omega \rho^n \int_{B_0^1} F(v_\varepsilon).$$

Assuming concentration at some point $x$ Theorem 12.3 yields the opposite inequality

$$\int_\Omega F(u_\varepsilon) \;\leq\; (\rho(x) + o(1))^n \int_{B_0^1} F(v_\varepsilon).$$

Hence $x$ must be a n-harmonic center. However, we do not know how to prove single point concentration for general integrands. For the volume integrand in two dimensions the above result has been obtained in [49].

**Theorem 12.13 (Conformal volume integrand)** *Let $\Omega$ be a bounded domain in $\mathbb{R}^n$ satisfying the exterior ball condition. If $p = n = 2$ (probably any $n$) then the extremal sets $A_\varepsilon$ for*

$$\sup\{|A| \,:\, A \subset \Omega,\ \mathrm{ncap}_\Omega(A) \leq \varepsilon^n\}$$

*concentrate at a n-harmonic center of $\Omega$ and*

$$|A_\varepsilon| \;=\; |B_0^1| \left(\max_\Omega \rho^n + o(1)\right) \exp\left(-\alpha_n \varepsilon^{-\frac{n}{n-1}}\right) \tag{12.2}$$

*as $\varepsilon \to 0$. Small balls centered at a n-harmonic center are nearly optimal in the sense that they satisfy the same asymptotic formula.*

*Proof.* Comparison with small balls centered at a n-harmonic center and the formula for the n-capacity of small balls (Theorem 9.2) yields the lower bound in (12.2). Equivalently we could use n-harmonic transplantation for this purpose.

For the upper bound denote by $u_\varepsilon$ the n-capacity potential of $A_\varepsilon$ and $v_\varepsilon := u_\varepsilon/\varepsilon$ with $\|\nabla v_\varepsilon\|_n \leq 1$. For a subsequence we have $v_\varepsilon \rightharpoonup v$ weakly in $D^{1,n}(\Omega)$. We use the Minty-Browder method (see Evans [46, p. 52]) to show that the weak limit is $v = 0$. This method only requires monotonicity of the operator involved. The n-Laplacian is monotone. With

$$E(u) \quad := \quad |\nabla u|^{n-2} \nabla u$$

we have

$$\begin{aligned} & (E(U) - E(V)) \cdot (\nabla U - \nabla V) \\ \geq \; & \frac{1}{2}\left(|\nabla U|^{n-2} + |\nabla V|^{n-2}\right) |\nabla U - \nabla V|^2 \;\; \geq \;\; 0. \end{aligned}$$

We apply this relation to $V = v_\varepsilon$ and a smooth perturbation $U = v + t\eta$, $t > 0$, of the weak limit. This yields

$$\begin{aligned} 0 \;\; \leq \;\; & \int_\Omega E(v + t\eta) \cdot \nabla(v - v_\varepsilon) + t \int_\Omega E(v + t\eta) \cdot \nabla\eta \\ & - \int_\Omega E(v_\varepsilon) \cdot \nabla(v - v_\varepsilon) - t \int_\Omega E(v_\varepsilon) \cdot \nabla\eta. \end{aligned}$$

In the limit $\varepsilon \to 0$ the first term vanishes by weak convergence of $(v_\varepsilon)$. So does the third one due to boundedness of $(E(v_\varepsilon))$ in the dual space of $L^n(\Omega)$ and the last one by the argument already presented in Example 4.1. We are left with the second term. Division by $t$ and $t \to 0$ leads to

$$0 \;\; \leq \;\; \int_\Omega E(v) \cdot \nabla\eta$$

for every test function $\eta$. Replacing $\eta$ by $-\eta$ yields the opposite inequality and $E(v) = 0$ in a weak sense. Hence $v = 0$ as desired.

This implies concentration of the extremal sets $A_\varepsilon$. To see this it suffices to exclude compactness for the normalized potentials $(v_\varepsilon)$. We choose a subsequence for which $|\nabla u_\varepsilon|^n \overset{*}{\rightharpoonup} \mu$. By the conformal concentration compactness alternative it suffices to show that

$$\int_\Omega \left(\exp\left(\alpha_n |v_\varepsilon|^{\frac{n}{n-1}}\right) - 1\right) \;\; \not\to \;\; \int_\Omega \left(\exp\left(\alpha_n |v|^{\frac{n}{n-1}}\right) - 1\right) \;\; = \;\; 0$$

Using $v_\varepsilon = \varepsilon^{-1}$ on $A_\varepsilon$ and the lower bound already obtained we can estimate

$$\begin{aligned} \int_\Omega \left(\exp\left(\alpha_n |v_\varepsilon|^{\frac{n}{n-1}}\right) - 1\right) \;\; & \geq \;\; |A_\varepsilon| \left(\exp\left(\alpha_n \varepsilon^{-\frac{n}{n-1}}\right) - 1\right) \\ & \geq \;\; |B_0^1| \left(\max_\Omega \rho^n + o(1)\right). \end{aligned}$$

Thus the left hand side cannot tend to 0 and the sets $A_\varepsilon$ concentrate at a single point. We are left to show that the concentration point is a n-harmonic center of

$\Omega$ and that (12.2) holds. Assume $\Omega \subset B_0^1$. By the isoperimetric inequality (2.10) and the change of domain formula (Theorem 9.5) we can estimate

$$\begin{aligned} |A_\varepsilon| &\leq \left|B_0^1\right| \exp\left(-\alpha_n \mathrm{nmod}_{B_0^1}(A_\varepsilon)\right) \\ &= \left|B_0^1\right| \exp\left(-\alpha_n(\mathrm{nmod}_\Omega(A_\varepsilon) + \tau(x) + o(1))\right) \\ &= \left|B_0^1\right| \left(\rho(x)^n + o(1)\right) \exp\left(-\alpha_n \varepsilon^{-\frac{n}{n-1}}\right) \end{aligned}$$

by (8.3). Together with the lower bound we conclude that $x$ must be a n-harmonic center. □

Another example for which we can prove concentration at a n-harmonic center is the critical plasma integrand. If $\Omega$ is not all of $\mathbb{R}^n$ concentration at some point follows from the concentration theorem (Theorem 7.3). For the linear case ($q = 2$) in two dimensions concentration of the plasma cores at a harmonic center has been proved by Caffarelli and Friedman [30]. Similar and more detailed results for $q \geq 2$ have been derived with C. Bandle [17] and J. Wei [59].

# Chapter 13

# Applications

In the applications of this chapter concentration of low energy extremals is already built into the problem. Thus we are left to identify the concentration point and to provide asymptotic formulas for the energy. These informations are important for the technical applications.

## 13.1 Optimal location of a small spherical conductor

Suppose we want to determine the optimal location of a pipeline for a hot liquid within a tube of given shape such that the heat leakage to the surrounding ground is minimal (Figure 13.1). Minimization of the current leakage between the two conductors of a coaxial cable leads to the same mathematical problem. We assume that the tube is filled with a homogeneous, isotropic insulating material and that the pipeline of cross section $B_x^r$ is at temperature $u = 1$ while the ground outside the tube $\Omega$ has temperature $u = 0$. By Fourier's law the heat flux depends linearly on the temperature gradient $j = -a\nabla T$. In the case of an isotropic and homogeneous material the heat conductance $a$ is a constant multiple of the identity matrix. Throughout the insulation layer the heat flux is source free. Thus the temperature is harmonic in this range. By Green's identity the rate of heat leakage through the boundary

$$\int_{\partial\Omega} j \cdot \nu \quad = \quad a(T(\partial B_x^r) - T(\partial\Omega)) \operatorname*{cap}_{\Omega}(B_x^r)$$

is proportional to the capacity of the conductor with respect to the surrounding ground. Given the exterior shape of the insulation layer the optimal location of the conductor $B_x^r$ is thus obtained by minimization of

$$E_r(x) \quad := \quad \operatorname*{cap}_{\Omega}(B_x^r).$$

with respect to $x$. Figure 13.2 shows how the heat loss depends on the location of the pipeline in the case of a special geometry. By the formula for the capacity

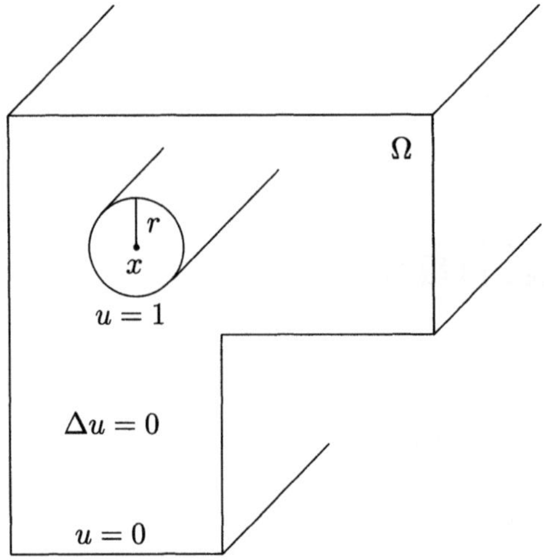


Figure 13.1: Pipeline within tube.

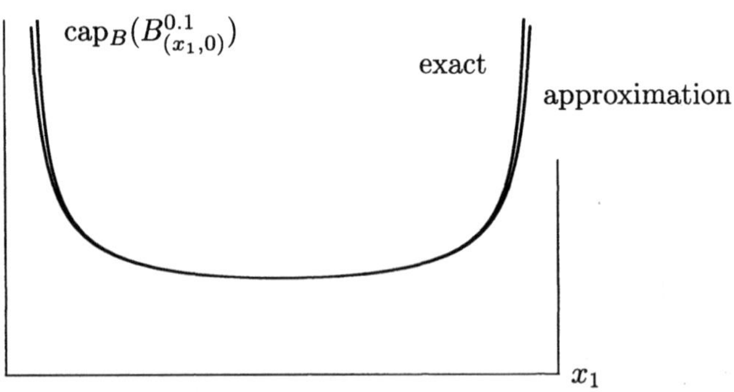


Figure 13.2: Dependence of the capacity of a small disk on its location.

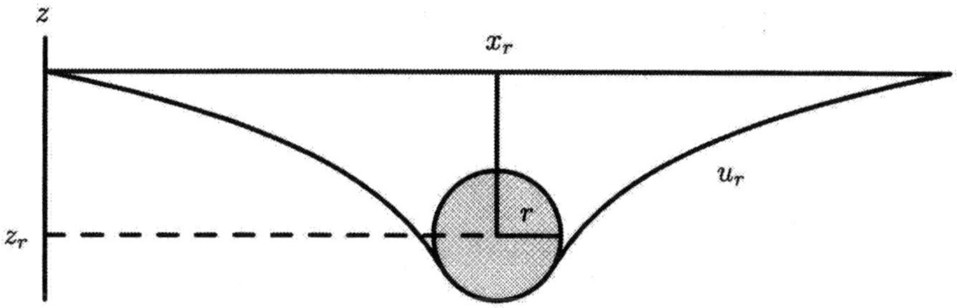


Figure 13.3: Ball on membrane.

of small balls (Theorem 9.2) we have

$$E_r(x) = \frac{1}{K(r) - \inf \tau} + \frac{\tau(x) - \inf \tau}{K^2(r)} + O\left(\frac{r}{K^2(r)}\right)$$

with the Robin function $\tau(x)$ playing the role of a renormalized energy. Thus for a point $x_r$ minimizing $E_r$ the second term must be of the same order as the error and we obtain

$$\begin{aligned} E_r(x_r) &= \frac{1}{K(r) - \inf \tau} + O\left(\frac{r}{K^2(r)}\right), \\ \tau(x_r) &= \inf \tau + o(1). \end{aligned}$$

Moreover $(x_r)$ has a subsequence that tends to a harmonic center $z$ of $\Omega$. If $z$ is a non-degenerate minimum of $\tau$, then $|x_r - z| = O(\sqrt{r})$. On unbounded domains $\inf \tau$ is not necessarily attained, instead $\tau(x_r) \to \inf \tau$. The existence of an optimal position of a conductor of arbitrary given shape has been proved by Colesanti [34]. In Chapter 14 we will also optimize the shape of the conductor.

## 13.2 Restpoints on an elastic membrane

Suppose we put a little ball on a horizontally stretched membrane and wait until it stops rolling. The question is: Where does it stop? The energy to be minimized is

$$\int_\Omega \sigma \sqrt{1 + |\nabla u|^2} + mgz$$

where $u$ denotes the deflection of the membrane, $\sigma$ its tension, and $mg$ the force of gravity (Figure 13.3). This functional has the following defect. The minimal energy among parametric surfaces supporting the ball is always $-\infty$. Only for large balls a non-parametric minimizer exists. Therefore we minimize the energy corresponding to the linearized operator, namely

$$E_r(u) = \frac{1}{2} \int_\Omega |\nabla u|^2 + Mz$$

with $M = mg/\sigma$. The ball is described by the obstacle

$$\psi^r_{x,z}(y) = \begin{cases} z - \sqrt{r^2 - |y-x|^2} & (y \in B^r_x), \\ 0 & (\text{otherwise}) \end{cases}$$

and we seek to minimize $E_r(u)$ among all $u \in D^{1,2}(\Omega)$ subject to $u \le \psi^r_{x,z}$ with the free parameters $x \in \Omega$ and $z \in \mathbb{R}$. For fixed $r$, $x$, $z$ the existence of a minimizing $u_r$ follows from the general theory of variational inequalities. In order to estimate the energy we pinch the ball between two artificial obstacles, namely

$$(z - r) \le u_r \le (z + r) \text{ on } B^r_x.$$

These relations carry over to the elastic energy of the membranes spanned over these obstacles

$$(z-r)^2 \operatorname*{cap}_\Omega(B^r_x) \le \int_\Omega |\nabla u_r|^2 \le (z+r)^2 \operatorname*{cap}_\Omega(B^r_x).$$

Therefore

$$\begin{aligned} E_r(x,z) &= \frac{1}{2} z^2 \operatorname*{cap}_\Omega(B^r_x) + Mz + O(rz \operatorname*{cap}_\Omega(B^r_x)) \\ &= -\frac{M^2}{2\,\mathrm{cap}_\Omega(B^r_x)} + \frac{\mathrm{cap}_\Omega(B^r_x)}{2}\left(z + \frac{M}{\mathrm{cap}_\Omega(B^r_x)}\right)^2 + O(rz \operatorname*{cap}_\Omega(B^r_x)). \end{aligned}$$

The minimum over $z$ is attained at

$$z_r = -\frac{M}{\mathrm{cap}_\Omega(B^r_x)} + O\left(\sqrt{r\,|z_r|}\right)$$

because for such $z$ the quadratic term is of the same order as the error. The fact that $z_r$ itself appears in the error is harmless because by comparison with radially symmetric solutions we can show a priori that $z_r \to -\infty$ as $r \to 0$. The minimal energy is given by

$$E_r(x) = -\frac{M^2}{2\,\mathrm{cap}_\Omega(B^r_x)} + O(r) = \frac{M^2}{4\pi}(\log r + \tau(x)) + O(r)$$

according to the formula for the capacity of small balls (Theorem 9.2). The minimum over $x$ is attained at a point $x_r$ with $\tau(x_r) = \inf \tau + O(r)$. Hence the altitude of the ball at rest and the energy of this least energy configuration is given by

$$\begin{aligned} z_r &= \frac{M}{2\pi}(\log r + \inf \tau) + O(\sqrt{r\,|\log r|}), \\ E &= \frac{M^2}{4\pi}(\log r + \inf \tau) + O(r). \end{aligned}$$

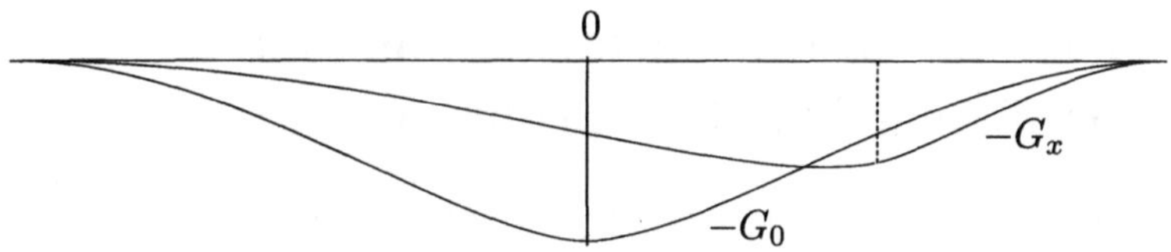


Figure 13.4: Deflection of circular clamped plate due to point load.

In particular the accumulation points of $(x_r)$ are harmonic centers of $\Omega$ and $|x_r - z| = O(\sqrt{r})$ if $(x_r)$ tends to a non-degenerate harmonic center $z$. Convergence to a harmonic center has been obtained by Elliott and Friedman [45] using a different argument. The above estimate for the distance to the harmonic center is sharper.

## 13.3 Restpoints on an elastic plate

The normal displacement of a clamped plate supporting the load $f$ is given by the solution of

$$\begin{aligned} \Delta^2 w &= f \text{ in } \Omega, \\ w &= 0 \text{ on } \partial\Omega, \\ \partial_\nu w &= 0 \text{ on } \partial\Omega. \end{aligned}$$

Suppose we seek to find the location of a point load moving freely on a horizonatal elastic plate (Figure 13.4). The mathematical problem is to minimize the sum of elastic energy plus potential energy

$$E = \frac{1}{2}\int_\Omega (\Delta w)^2 + Mz$$

among all positions of the point load $(x, z) \subset \Omega \times \mathbb{R}$ and all possible deflections of the plate $w \in H_0^{2,2}(\Omega)$ subject to $w(x) \leq z$. For fixed $x$ and $z$ the deflection of the plate is given by a multiple of the biharmonic Green's function (Section 8.5.2)

$$w = \frac{z}{G_x(x)} G_x.$$

with

$$E = -\frac{z^2}{2\tau(x)} + Mz$$

because $\int_\Omega (\Delta G_x)^2 = G_x(x) = -\tau(x)$ by Green's identity. The minimizing vertical position of the load is $z = M\tau(x)$ leading to

$$E = \frac{M^2}{2}\tau(x).$$

Thus the minimum of the energy is achieved at a biharmonic center.

## 13.4 Location of concentration points in a semilinear Dirichlet problem

With J. Wei [58] we have studied the asymptotic properties of the positive solutions $u_q$ of

$$\begin{aligned} \Delta u + u^{q-1} &= 0 \text{ in } \Omega, \\ u &= 0 \text{ on } \partial\Omega \end{aligned} \tag{13.1}$$

as $q$ tends to the critical Sobolev exponent. Brézis and Peletier [28] conjectured that for $n \geq 3$ the maximum points of these solutions accumulate at a critical point of the Robin function. This has been confirmed by Rey [108] and Han [67]. In [58] we have shown that the same conclusion holds in two dimensions as $q$ tends to infinity and we presented a variational approach that covers all dimensions in a unified way. As a byproduct it turned out that the concentration point for variational solutions is a harmonic center. As already mentioned in the introduction Bahri, Li, and Rey [14] have identified multiple concentration points of arbitrary positive solutions of this problem.

Let $u_q$ denote a solution of the variational problem

$$S_q(\Omega) := \sup \left\{ \frac{\int_\Omega |u|^q}{\left(\int_\Omega |\nabla u|^2\right)^{q/2}} : u \in D^{1,2}(\Omega),\ u \not\equiv 0 \right\}, \tag{13.2}$$

i.e. an extremal function for the corresponding Sobolev inequality. We consider the range $2 < q < 2^*$. In two dimensions $2^* = \infty$. For $n \geq 3$ we have $S_q(\Omega) \to S^*$ as $q \to 2^*$. In two dimensions $S_q(\Omega) \to \infty$ as $q \to \infty$. After scaling we can assume that $u_q$ is a positive solution of the Euler Lagrange equation (13.1).

**Theorem 13.1 ([58])** *Let $\Omega$ be a bounded domain of dimension $n \geq 2$ satisfying the exterior ball condition. For $2 < q < 2^*$ let $x_q$ be a maximum point of the solution $u_q$ of (13.2) satisfying (13.1). Then*

$$\lim_{q \to 2^*} \rho(x_q) = \max_\Omega \rho.$$

*In particular every accumulation point of $(x_q)$ is a harmonic center of $\Omega$. Moreover*

$$S_q(\Omega) \geq S_q(B_0^{\sup \rho}) = \sup \rho^{n_q} S_q(B_0^1)$$

*where $n_q := n - \frac{q}{2}(n-2)$.*

In two dimensions we have $n_q = 2$ while in higher dimensions $n_q = \frac{n-2}{2}(2^* - p) \to 0$ as $q \to 2^*$. Under dilations $S_q(\Omega)$ scales as

$$S_q(s\Omega) = s^{n_q} S_q(\Omega). \tag{13.3}$$

In fact, with $v_q(x) := u_q(x/s)$ we have

$$S_q(s\Omega) \geq \frac{\int_\Omega |v_q|^q}{\left(\int_\Omega |\nabla v_q|^2\right)^{q/2}} = \frac{s^n \int_\Omega |u_q|^q}{\left(s^{n-2} \int_\Omega |\nabla u_q|^2\right)^{q/2}} = s^{n_q} S_q(\Omega).$$

The reverse scaling with an extremal for $S_q(s\Omega)$ yields the opposite inequality. The bound for $S_q(\Omega)$ is obtained by harmonic transplantation. Preliminary results in two dimensions using different methods have been obtained by Ren and Wei [105, 106].

# Chapter 14

# Bernoulli's Free-boundary Problem

As already mentioned in the introduction Bernoulli's free-boundary problem arises in ideal fluid dynamics, optimal design, electro chemistry, electro statics, and further applications. In the *interior Bernoulli problem* a connected domain $\Omega$ in $\mathbb{R}^n$ and a constant $Q > 0$ are given. The task is to find a subset $A \subset \Omega$ and a potential $u : \Omega \setminus A \to \mathbb{R}$ such that

$$\begin{aligned} -\Delta u &= 0 \quad \text{in } \Omega \setminus A, \\ u &= 0 \ \text{ on } \partial\Omega, \\ u &= 1 \ \text{ on } \partial A, \\ \frac{\partial u}{\partial \nu} &= Q \ \text{ on } \partial A \end{aligned}$$

(see Figure 1.1 on page 2). In the *exterior Bernoulli problem* $\partial A$ is exterior to $\partial\Omega$ with $u = 1$ on $\partial\Omega$, $u = 0$ on $\partial A$ and $Q < 0$ (Figure 14.1). The same problem can be posed for the p-Laplacian (Acker and Meyer [3]). We mainly consider the semilinear problem. Typically the interior Bernoulli problem has two solutions, an elliptic one close to the fixed boundary, and a hyperbolic (low energy) solution far from the boundary. Hyperbolic solutions are more delicate for analysis and numerical approximation. Nevertheless there is a second order trial free-boundary method, the implicit Neumann scheme (Section 14.3), with equally good performance for both types of solutions. See our paper with M. Rumpf [56] for the convergence proof. To begin with we describe the applications in physics and industry in detail.

**Free surfaces in ideal fluid dynamics** Consider an inviscid, incompressible, irrotational, stationary, horizontal flow of velocity $v$ with stream function $u$. We assume that the fluid circulates around a bubble of air denoted by $A$. At the free boundary $\partial A$ the normal stress of the liquid coincides with that of the air.

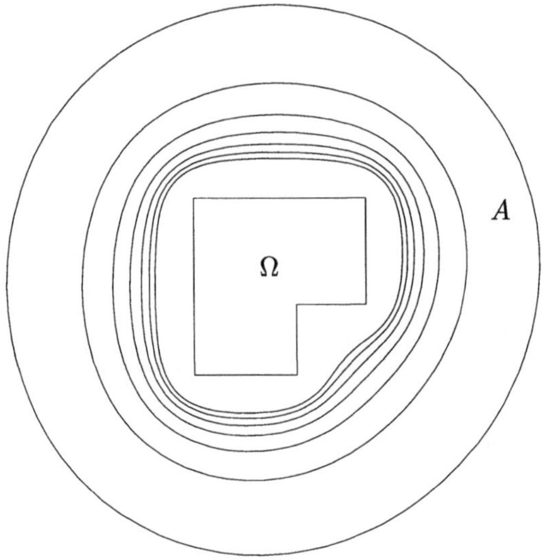


Figure 14.1: Solutions of exterior Bernoulli problem.

Thus *Bernoulli's law* (named after Daniel Bernoulli, 1700-1782)

$$\frac{1}{2}|v|^2 + \frac{p}{\rho} + gz = \text{const.}$$

reduces to $|v| = |\nabla u| = \text{const}$. Moreover the potential is constant along the streamlines $\partial\Omega$ and $\partial A$. The equipotential lines in Figure 1.1 can be interpreted as streamlines of such a flow. Jet flow and wakes are specific configurations considered by Friedrichs [62], Friedman [61], and Tepper [126]. As the pressure of the air decreases the vortex core shrinks to a stationary point vortex (Chapter 15).

**Optimal insulation** Bernoulli's free-boundary problem also arises in certain optimal design problems. Suppose we want to design an insulation layer that minimizes the heat or current leakage subject to a given amount of insulating material. Given the inner or outer geometry of the insulation layer the optimal shape of the conductor $A$ is obtained by minimization of the capacity among sets of equal volume. In Section 13.1 we already found the optimal location of a spherical conductor. Now we also optimize its shape. We will see that the optimal set $A$ is a solution of Bernoulli's free-boundary problem. Specific applications are minimization of the current leakage from insulated wires and coaxial cables (Acker [1]) or minimization of the capacity of condensers and resistors.

**Electrolytic drilling and galvanization** The same free-boundary problem arises in electro chemical machining (Lacy-Shillor [79], Crank [35, p. 85]). In this application $A \setminus \Omega$ describes an electrolytic cell bounded by two electrodes. One of them is dissolved by the electrolyte. The region occupied by the electrolyte grows at a rate which depends on the magnitude of the electric field at the surface of the electrode. This leads to the following moving boundary problem. Given the drill $\Omega$ (cathode), one tries to determine the shape of the workpiece $\mathbb{R}^n \setminus A(t)$ (anode) at time $t$ and the electrostatic potential $u(t)$, such that

$$\begin{aligned} -\Delta u(t) &= 0 \text{ in } A(t) \setminus \Omega, \\ u(t) &= 0 \text{ on } \partial A(t), \\ u(t) &= 1 \text{ on } \partial \Omega, \\ \dot{x} \cdot \nu &= \varphi(|\nabla u|) \quad \text{on } \partial A(t). \end{aligned}$$

Typically the machining rate $\varphi$ vanishes below a certain threshold current $Q$ and $\varphi(q) > 0$ for $q > Q$. If initially $|\nabla u| < Q$ on $\partial A$ then $u(t)$ tends to a stationary solution which is a solution of the exterior Bernoulli problem (Figure 14.1). For general $\varphi$ the evolutionary problem is unsolved (Fasano [47]). Strictly speaking this is an inverse Bernoulli problem because in the application the shape of the workpiece is prescribed and the tool is to be designed. Galvanization leads to the inverse interior Bernoulli problem. In order to produce a layer of constant thickness it is necessary to a apply an electric field whose magnitude is constant on the surface of the workpiece.

## 14.1 Variational methods for Bernoulli's free-boundary problem

We describe two variational methods to produce solutions of Bernoulli's free-boundary problem. Depending on the physical problem we choose a constrained or an unconstrained approach. In the optimal design problem the volume of $A$ or $\Omega \setminus A$ is given and $Q$ enters as a Lagrange multiplier. In the other applications $Q$ is given. For these the unconstrained formulation is more adequate. The fact that Bernoulli's free-boundary problem is the Euler Lagrange equation for these variational problems is a consequence of the subsequent formula for the variation of the p-capacity subject to regular variations of the boundary. In the semilinear case it is due to Poincaré [102]. It is a special case of Hadamard's domain derivative formula used in shape optimization (see e.g. Simon [118] or Sokolowski and Zolesio [119]).

**Lemma 14.1 (Regular variation of p-capacity)** *The first variation of the p-capacity of a set $A$ subject to a variation $\delta x$ of the boundary $\partial(\Omega \setminus A)$ is given by*

$$\delta \mathrm{pcap}_\Omega(A) = -(p-1) \int_{\partial(\Omega \setminus A)} |\nabla u|^p \, \delta x \cdot \nu,$$

*where $u$ denotes the p-capacity potential of $A$ with respect to $\Omega$.*

*Proof.* Let $u^\delta = u + \delta u$ be the p-capacity potential of $A^\delta$ with respect to $\Omega^\delta$ where $\partial(\Omega \setminus A)^\delta = \partial(\Omega \setminus A) + \delta x$. The correction $\delta u$ solves the linear problem

$$\begin{aligned} -\mathrm{div}\,(|\nabla u|^{p-2}\,\nabla \delta u) &= 0 \text{ in } \Omega \setminus A, \\ \delta u &= -\nabla u \cdot \delta x \text{ on } \partial(\Omega \setminus A). \end{aligned}$$

Therefore the perturbed p-capacity is (up to higher order terms)

$$\begin{aligned} &\int_{\Omega^\delta \setminus A^\delta} |\nabla u^\delta|^p \\ = &\int_{\Omega^\delta \setminus A^\delta} |\nabla u|^p + p \int_{\Omega^\delta \setminus A^\delta} |\nabla u|^{p-2}\,\nabla u \cdot \nabla \delta u \\ = &\int_{\Omega \setminus A} |\nabla u|^p + \int_{\partial(\Omega \setminus A)} |\nabla u|^p\,\delta x \cdot \nu + p \int_{\partial(\Omega \setminus A)} |\nabla u|^{p-2}\,\nabla u \cdot \nu\,\delta u. \end{aligned}$$

On $\partial(\Omega \setminus A)$ we have $\nabla u \cdot \nu \delta u = -(\nabla u \cdot \nu)(\nabla u \cdot \delta x) = |\nabla u|^2\,\delta x \cdot \nu$ because $\partial(\Omega \setminus A)$ is a level set of $u$. □

In the constrained approach we maximize the volume among sets of equal capacity (3.7). The variation of the volume is given by $\delta\,|A| = -\int_{\partial A} \delta x \cdot \nu$. By the Lagrange multiplier rule solutions of the constrained variational problem (3.7) are weak solutions of Bernoulli's free-boundary problem. The number $Q$ plays the role of a Lagrange multiplier. The constrained formulation has the advantage that it always admits a nontrivial solution. Existence of maximizers follows from upper semi continuity of the volume integrand. An unconstrained method has been used by Alt and Caffarelli [7]. They minimize

$$J_Q(u) := \int_\Omega |\nabla u|^p + \int_{\{u<1\}} Q^2$$

among all $u \in D^{1,2}(\Omega)$. A nontrivial minimizer exists if and only if $\inf J_Q < J_Q(0) = Q^2\,|\Omega|$. Minimizers are always elliptic solutions in the sense of Section 14.2 below. The penalization method used by Aguilera, Alt, and Caffarelli [6] is equivalent to the constrained variational method. They minimize

$$J_\alpha^\varepsilon(u) := \int_\Omega |\nabla u|^2 + f^\varepsilon(|\{u < 1\}| - \alpha)$$

over $D^{1,2}(\Omega)$ where

$$f^\varepsilon(\tau) := \begin{cases} \frac{\tau}{\varepsilon} & (\tau \geq 0), \\ \varepsilon\tau & (\tau < 0). \end{cases}$$

Surprisingly the minimizers of $J_\alpha^\varepsilon$ satisfy the exact constraint $|\{u < 1\}| = \alpha$ already for small $\varepsilon$, not only in the limit $\varepsilon \to 0$.

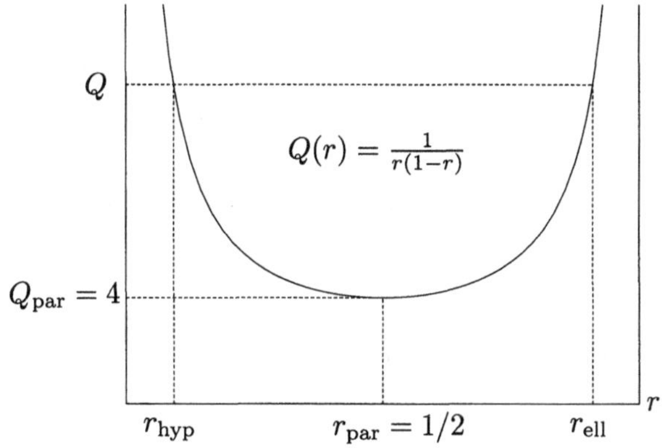


Figure 14.2: Slope of radial solutions at the free boundary in three dimensions.

## 14.2 Non-degenerate, elliptic, hyperbolic solutions

Several relevant features of Bernoulli's free-boundary problem are already manifest for radial solutions. For a ball $\Omega = B_0^R$ every interior solution is a concentric ball $A = B_0^r$ and the potential $u$ is given by (2.6). The slope of this potential at the free boundary is

$$Q(r,R) = \begin{cases} -\frac{1}{r\log(r/R)} & (n=2), \\ \frac{n-2}{r(1-(r/R)^{n-2})} & (n \geq 3). \end{cases}$$

The same formula is valid for the exterior problem. Figure 14.2 shows the function $r \mapsto Q(r,1)$ for the unit ball in three dimensions. Three different types of solutions occur:

1. Elliptic solutions which increase as $Q$ increases,
2. Hyperbolic solutions which decrease as $Q$ increases,
3. A degenerate or parabolic or solution corresponding to the minimal value $Q_{\text{par}}(\Omega)$.

This terminology is due to Beurling [22]. He made the fundamental observation that elliptic solutions in two dimensions can be constructed by the method of sub and super solutions. For hyperbolic solutions this method fails. We extend his classification to general domains by linearizing the solutions of Bernoulli's free-boundary problem with respect to the data. In particular we define a concept of non-degenerate solutions that is crucial for local well-posedness of Bernoulli's free-boundary problem and convergence of trial free-boundary methods. For this purpose we consider a smooth solution $(A, u)$ of the p-Bernoulli problem and a solution of the perturbed problem

$$\begin{aligned} -\mathrm{div}\left(\left|\nabla u^\delta\right|^{p-2} \nabla u^\delta\right) &= 0 \text{ in } \Omega \setminus A^\delta, \\ u^\delta &= 0 \text{ on } \partial\Omega, \\ u^\delta &= 1+\delta v \quad \text{on } \partial A^\delta, \\ \frac{\partial u^\delta}{\partial \nu} &= Q+\delta Q \quad \text{on } \partial A^\delta. \end{aligned}$$

with regular variations $\delta v$ and $\delta Q$ of the Dirichlet and Neumann data $(1, Q)$ on $\partial A$ and $\partial A^\delta = \partial A + \delta\nu$, $u^\delta = u + \delta u$.

**Theorem 14.2 (Linearization of p-Bernoulli problem)** *The first variation $\delta u$ of a solution of the p-Bernoulli problem subject to a regular variation $(\delta v, \delta Q)$ of the Dirichlet and Neumann data on $\partial A$ is given by the solution of the linearized problem*

$$\begin{aligned} -div\,(|\nabla u|^{p-2} \nabla \delta u) &= 0 \;\; in \;\; \Omega \setminus A, \\ \delta u &= 0 \;\; on \;\; \partial\Omega, \\ \frac{\partial \delta u}{\partial \nu} + \frac{n-1}{p-1} H \delta u &= \delta R \quad on \;\; \partial A, \\ \delta R &:= \delta Q + \frac{n-1}{p-1} H\, \delta v. \end{aligned} \tag{14.1}$$

*where $H$ denotes the mean curvature of $\partial A$ with respect to $\nu$. The variation of $\partial A$ is given by*

$$\delta\nu = \frac{\delta v - \delta u}{Q} \nu$$

*evaluated on $\partial A$.*

*Proof.* Define the function $s$ on $\partial A$ by $\delta\nu = s\nu$. We expand the unperturbed potential and its normal derivative at the free boundary to first order:

$$\begin{aligned} u(x+\delta\nu) &= 1 + Qs, \\ \frac{\partial u}{\partial \nu}(x+\delta\nu) &= Q + \frac{\partial^2 u}{\partial \nu^2}(x) s. \end{aligned}$$

Using the equation for $u$ we can calculate the second derivative. The mean curvature of a level set of $u$ is given by the equation $\mathrm{div}\, \frac{\nabla u}{|\nabla u|} = (n-1)H$. Using this fact we can expand the p-Laplacian near the free boundary as

$$\begin{aligned} \mathrm{div}\,(|\nabla u|^{p-2} \nabla u) &= \mathrm{div}\left(|\nabla u|^{p-1} \frac{\nabla u}{|\nabla u|}\right) \\ &= (p-1)\, |\nabla u|^{p-2} \frac{\partial^2 u}{\partial \nu^2} + (n-1) H\, |\nabla u|^{p-2} \frac{\partial u}{\partial \nu}. \end{aligned}$$

Thus the second the derivative of the potential is

$$\frac{\partial^2 u}{\partial \nu^2}(x) = -\frac{n-1}{p-1}H(x)Q$$

such that

$$\frac{\partial u}{\partial \nu}(x+\delta\nu) = Q - \frac{n-1}{p-1}H(x)\left(u(x+\delta\nu)-1\right).$$

The Neumann condition for the perturbed problem is transformed into

$$\begin{aligned} Q+\delta Q &= \frac{\partial u^\delta}{\partial \nu} = Q - \frac{n-1}{p-1}H(u-1) + \frac{\partial \delta u}{\partial \nu} \\ &= Q - \frac{n-1}{p-1}H(\delta v - \delta u) + \frac{\partial \delta u}{\partial \nu} \quad \text{on } \partial A^\delta. \end{aligned}$$

This is the second boundary condition of (14.1). From the Dirichlet condition for $u^\delta$ we obtain the formula for the normal correction of the free boundary. □

In the semilinear case the weak form of (14.1) is

$$\int_{\Omega\setminus A} \nabla\delta u \cdot \nabla\eta + \int_{\partial A}(n-1)H\delta u\,\eta = \int_{\partial A}\delta R\,\eta$$

for all $\eta$ in the space

$$H^1_{\partial\Omega}(\Omega\setminus A) := \{\eta \in H^1(\Omega\setminus A) : \eta = 0 \text{ on } \partial\Omega\}.$$

If the operator

$$\begin{aligned} L_{A,H} &: H^1_{\partial\Omega}(\Omega\setminus A) \to H^1_{\partial\Omega}(\Omega\setminus A)^*, \\ (L_{A,H}\delta u)\eta &= \int_{\Omega\setminus A}\nabla\delta u\cdot\nabla\eta + \int_{\partial A}(n-1)H\delta u\,\eta \end{aligned}$$

is invertible then Bernoulli's free boundary problem is locally well posed. In this case an approximation $(A^\delta, u^\delta)$ of a smooth solution $(A, u)$ solves the perturbed problem with certain $(\delta v, \delta Q)$. In order to show that $L_{A,H}$ is invertible it suffices to show that it has trivial kernel. In fact, if $\partial A$ is $C^3$ then $L_{A,H}$ is a Fredholm Operator. The Fredholm property follows from classical elliptic theory (see e.g. Wloka [131, Theorem 13.1 and Example 11.3]). In our case we can argue as follows. With $H = 0$ the inverse of $L_{A,0}$ is the solution operator of the standard Neumann problem with $\delta u = 0$ on $\partial\Omega$ and $\frac{\partial \delta u}{\partial \nu} = \delta R$ on $\partial A$ and thus exists. Due to the embedding $H\delta u \in H^{1/2}(\partial A) \subset\subset H^1_{\partial\Omega}(\Omega\setminus A)^*$ for smooth $H$ and every $\delta u \in H^1_{\partial\Omega}(\Omega\setminus A)$ the difference $L_{A,H} - L_{A,0}$ is a compact operator. Thus $L_{A,H}$ is a Fredholm operator. This motivates our definition of non-degeneracy.

**Definition 14.3 (Non-degenerate solutions)** *We say that the set $A$ is non-degenerate if the linearized problem* (14.1) *with $\delta R = 0$ has only the trivial solution $\delta u = 0$, i.e.* $\ker(L_{A,H}) = \{0\}$.

A sufficient criterion for non-degeneracy is $|Q| + (n-1)H > 0$ on $\partial A$ (see [56] or Acker [2]).

**Example 14.4** In the radial semilinear case $\Omega = B_0^1$, $A = B_0^r$ the linearized problem (14.1) assumes the form

$$\begin{aligned} -\Delta\delta u &= 0 \text{ in } B_0^1 \setminus B_0^r, \\ \delta u &= 0 \text{ on } \partial B_0^1, \\ \frac{\partial \delta u}{\partial \nu} - \frac{n-1}{r}\delta u &= \delta R \text{ on } \partial B_0^r. \end{aligned}$$

For constant $\delta R$ the solution is given by

$$\delta u(x) = \begin{cases} -\frac{r\,\delta R}{1+\log(r)} \log|x| & (n=2), \\ \frac{r^2 \delta R}{2r-1}\left(\frac{1}{|x|} - 1\right) & (n=3). \end{cases}$$

Note that $\delta R > 0$ implies $\delta u > 0$ in the elliptic regime $r > r_{\text{par}}$ and $\delta u < 0$ in the hyperbolic regime $r < r_{\text{par}}$. In the parabolic case $r = r_{\text{par}}$ the operator $L_{A,H}$ has a one dimensional kernel spanned by the capacity potential of $B_0^{r_{\text{par}}}$. This is the only degenerate direction for the parabolic solution. We show that the elliptic and the hyperbolic solutions are non-degenerate. In two dimensions every harmonic function on $B_0^1 \setminus B_0^r$ with $\delta u = 0$ on $\partial B_0^1$ is of the form

$$\begin{aligned} &\delta u(r\cos\eta, r\sin\eta) \\ =\ &a_0 \log r + \sum_{k=1}^{\infty} a_k(r^k - r^{-k})\cos(k\eta) + b_k(r^k - r^{-k})\sin(k\eta). \end{aligned}$$

The boundary condition on $\partial B_0^r$ assumes the form

$$\begin{aligned} &-\frac{\partial \delta u}{\partial r}(r) - \frac{1}{r}\delta u(r) \\ =\ &-\frac{1}{r}\Bigg(a_0(1+\log r) + \sum_{k=1}^{\infty} a_k((k+1)r^k + (k-1)r^{-k})\cos(k\eta) \\ &+ \sum_{k=1}^{\infty} b_k((k+1)r^k + (k-1)r^{-k})\sin(k\eta)\Bigg) = 0 \text{ for every } \eta. \end{aligned}$$

Thus either $r = 1/e = r_{\text{par}}$ and $\delta u = a_0 \log r$ or else $\delta u = 0$. In higher dimensions we use a similar decomposition with respect to spherical harmonics. Thus the parabolic solution is the only degenerate solution of this example. In two dimensions the solution operator of the linearized problem is given by the following formulas. If

$$\delta R(r\cos\eta, r\sin\eta) = R_0 + \sum_{k=1}^{\infty} R_k\cos(k\eta) + S_k\sin(k\eta)$$

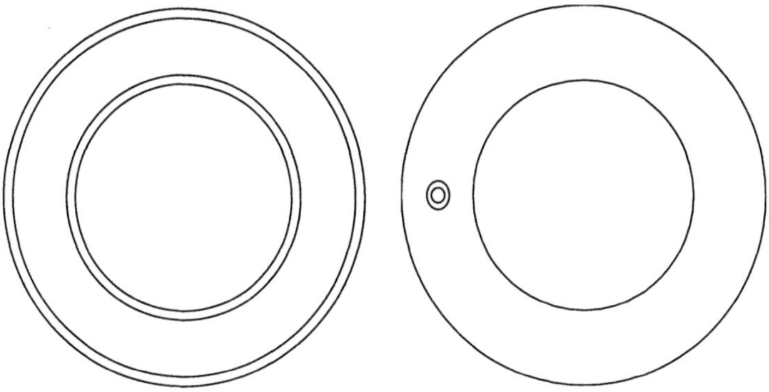

Figure 14.3: Domain with one non-degenerate elliptic and a one parameter family of degenerate hyperbolic solutions for the same value of $Q$.

then

$$a_0 = -\frac{r}{1+\log(r)} R_0, \tag{14.2}$$

$$a_k = -\frac{r}{(k+1)r^k + (k-1)r^{-k}} R_k \quad (k \geq 1) \tag{14.3}$$

and similarly for $b_k$. In particular the solution operator of the linearized problem is smoothing.

Figure 14.3 shows a different type of degeneracy. The domain is an annulus. The hyperbolic solutions are degenerate due to the symmetry of the domain. For each of them the operator $L_{A,H}$ has a one dimensional kernel. Now we are in position to give a precise classification of the solutions.

**Definition 14.5 (Elliptic, hyperbolic and parabolic solutions)** *We say that a connected component* $\Gamma$ *of* $\partial A$ *is an* elliptic (hyperbolic) solution *if the linearized problem* (14.1) *with* $\delta R = 1$ *is solvable and every solution satisfies*

$$\int_\Gamma \delta u > 0, \ (< 0)$$

*respectively. Otherwise* $\Gamma$ *is called* parabolic *or* degenerate.

Figure 14.4 illustrates this definition.

Note that elliptic (hyperbolic) solutions increase (decrease) in volume as $Q$ increases since the first variation of $|A|$ is given by $Q^{-1} \int_{\partial A} \delta u$. Parabolic solutions are typically degenerate.

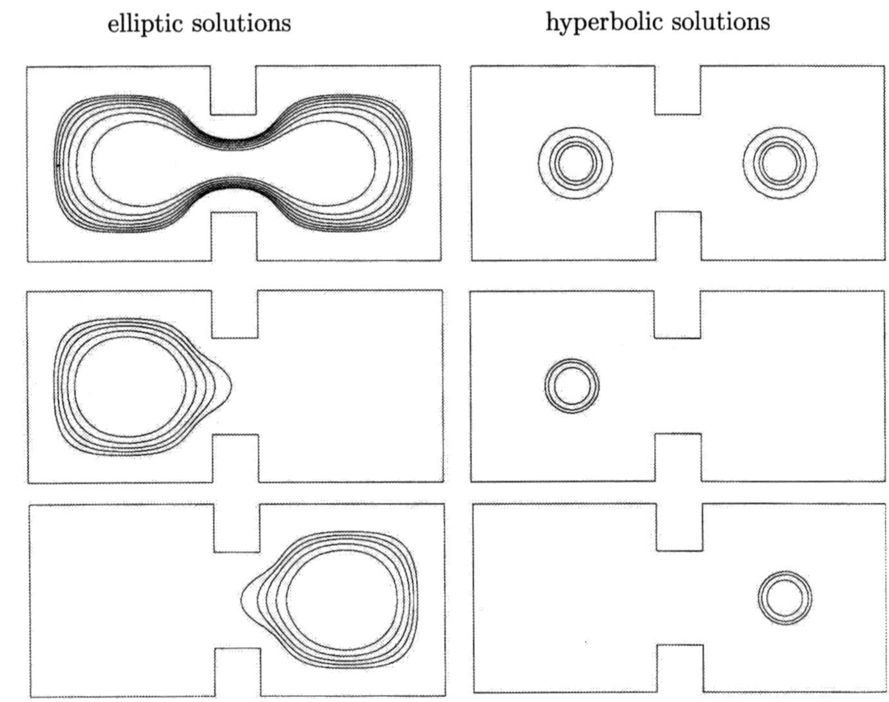


Figure 14.4: Domain with 3 elliptic and 3 hyperbolic solutions for the same value of $Q$.

## 14.3 Implicit Neumann scheme

The variational formula of Theorem 14.2 is equally useful for numerical purposes. It is the basis of the *implicit Neumann scheme* [56]. It is a second order trial free-boundary method. The figures of this chapter have been computed by this method. Starting from an initial guess for the free boundary we construct a sequence of approximations $\partial A_k$ of the free boundary. The idea is to impose the Neumann condition on the new approximation $\partial A_{k+1}$:

$$\frac{\partial u_k}{\partial \nu_{k+1}} = Q \text{ on } \partial A_{k+1}.$$

Using the variational formula this implicit condition can be replaced by an explicit one. We obtain the following scheme.

**Implicit Neumann method**

1. Choose an initial approximation $A_0$ of $A$.

2. For $k = 0, 1, \ldots$ proceed as follows:

(a) Solve

$$\begin{aligned} -\Delta u_k &= 0 \text{ on } \Omega \setminus A_k \\ u_k &= 0 \text{ on } \partial\Omega \\ \frac{\partial u_k}{\partial \nu_k} + (n-1)H_k(u_k - 1) &= Q \text{ on } \partial A_k \end{aligned}$$

with $H_k$ as in Theorem 14.2.

(b) Replace $\partial A_k$ by

$$\partial A_{k+1} := \partial A_k + \frac{1-u_k}{Q}\nu_k$$

with $u_k$ evaluated on $\partial A_k$.

3. Repeat from Step 2 until the accuracy goal is achieved.

We illustrate second order convergence of this method in the radial case. A proof for general domains is given in our paper with M. Rumpf [56]. On the unit ball $\Omega = B_0^1$ with $A = B_0^r$ the implicit Neumann method reduces to the iteration

$$r_k \mapsto r_{k+1} = \begin{cases} \frac{Qr_k - 1}{Q(1+\log(r_k))} & (n=2), \\ \frac{1/Q - r_k^2}{1-2r_k} & (n=3). \end{cases}$$

For $Q > Q_{\text{par}}$ the derivative of $r_k \mapsto r_{k+1}$ vanishes at the fixed points. Only for $Q = Q_{\text{par}}$ the scheme degenerates to a first order method.

## 14.4 Optimal shape of a small conductor

In Section 13.1 we determined the optimal location of a small conductor. Now we also optimize its shape. This means to determine the limit shape of small (hyperbolic) solutions of Bernoulli's free-boundary problem. By Corollary 7.13 we know that they are asymptotically spherical in the $L^1$-sense. In fact we expect a much stronger convergence to the spherical limit shape (Conjecture 14.6 below). By the Theorems 11.1 and 12.13 we also know that the concentration point is a harmonic center and that a small ball centered at a harmonic center provides an excellent approximate solution of the variational problem (3.7). This fact is illustrated by Figure 14.5. Beyond this we expect that a branch of non-degenerate hyperbolic solutions bifurcates from every non-degenerate local minimum of the Robin function. In this case we can also compute the asymptotic deviation from the spherical shape using the variational formula of Theorem 14.2.

**Conjecture 14.6** *Let $z$ be a non-degenerate local minimum of the Robin function. Then Bernoulli's free-boundary problem with*

$$Q = -\frac{K'(r)}{K(r) - \tau(z)}$$

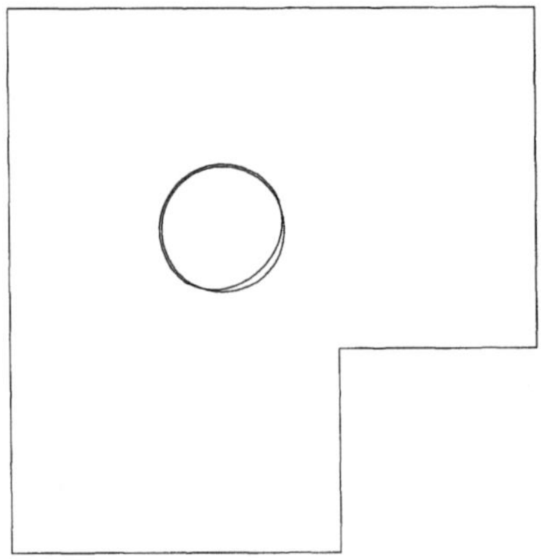

Figure 14.5: Comparison of hyperbolic solution with ball centered at a harmonic center.

*has a non-degenerate hyperbolic solution which is asymptotically given by*

$$\partial A^r = \partial B_z^r + \frac{1}{2}\frac{(x-z)^\top D^2 H_z(z)(x-z)}{rK'(r)}(x-z) + O(r^{n+2})$$

*for* $x \in \partial B_z^r$ *as* $r \to 0$.

The approximation of the free boundary involves a term of order $O(r)$ and one of order $O(r^{n+1})$. There are no terms of order $r^2, \dots, r^n$. By harmonicity of $H_z$ the approximate free boundary is an ellipsoid with half axes $r + r^{n+1}\lambda_k$ with $\sum_{k=1}^n \lambda_k = 0$. A careful look at Figure 14.3 shows that in fact the hyperbolic solution has the shape of an ellipse with the long axis in the direction predicted by the conjecture. We explain why the second term in the approximation of the free boundary must have the above noted form. We take $\partial B_z^r$ as approximate free boundary and

$$u := \frac{G_z}{K(r) - \tau(z)}$$

as approximate potential and apply the variational formula of Theorem 14.2. On $\partial B_z^r$ we have

$$\begin{aligned} G_z(x) &= K(r) - \tau(z) - \frac{1}{2}(x-z)^\top D^2 h_z(z)(x-z) + O(r^3), \\ \frac{\partial G_z}{\partial \nu}(x) &= -K'(r) - (x-z)^\top D^2 h_z(z)\nu + O(r^2) \end{aligned}$$

because $z$ is a critical point of the Robin function. Division by $K(r) - \tau(z)$ and $\nu = -(x-z)/r$ yields the following boundary data for the correction of the potential:

$$\begin{aligned}
\delta u(x) &= 1-\frac{1}{2}\frac{(x-z)^\top D^2H_z(z)(x-z)}{K(r)-\tau(z)}+O\left(\frac{r^3}{K(r)}\right),\\
\frac{\partial\delta u}{\partial\nu} &= Q+\frac{(x-z)^\top D^2H_z(z)(x-z)}{r(K(r)-\tau(z))}+O\left(\frac{r^2}{K(r)}\right),\\
r^{n+1}\delta R &= r^{n+1}(\delta Q+H\delta v)\\
&= \left(1+\frac{n-1}{2}\right)\frac{(x-z)^\top D^2H_z(z)(x-z)}{r(K(r)-\tau(z))}+O\left(\frac{r^2}{K(r)}\right).
\end{aligned}$$

For simplicity we continue with the two dimensional case where

$$D^2H_z(z) = \begin{pmatrix} h_{11} & h_{12}\\ h_{12} & -h_{11}\end{pmatrix},\quad x-z = (r\cos\eta, r\sin\eta).$$

Therefore

$$\begin{aligned}
(x-z)^\top D^2H_z(z)(x-z) &= r^2(h_{11}\cos(2\eta)+h_{12}\sin(2\eta)),\\
r^3\delta R &= \frac{3}{2}\frac{r(h_{11}\cos(2\eta)+h_{12}\sin(2\eta))}{K(r)-\tau(z)}+O\left(\frac{r^2}{K(r)}\right).
\end{aligned}$$

Here we use the notation of Theorem 14.2. The calculation for the radial case (14.2) indicates that the solution operator of the linearized problem for $A=B_z^r$ multiplies the k-th Fourier coefficient with a term of order $O(r^{k+1})$. We apply this to all coefficients. The result turns out to be inferior to $r^3\delta v$. Disregarding convergence questions we get

$$\begin{aligned}
r^3\delta u &= O\left(\frac{r^3}{K(r)}\right),\\
r^3\delta\nu\cdot\nu &= r^3\frac{\delta v-\delta u}{Q}\\
&= \frac{1}{2}\frac{(x-z)^\top D^2H_z(z)(x-z)}{K'(r)}+O\left(\frac{r^3}{K'(r)}\right).
\end{aligned}$$

This is the desired second term.

# Chapter 15

# Vortex Motion in Two Dimensional Hydrodynamics

The Robin function has an important application in planar fluid dynamics. Consider an incompressible and non-viscous fluid. We will see that the interaction of point vortices with the walls of the container is governed by the Robin function, while the interaction of vortex pairs is governed by the Green's function. As an application of the properties of the Robin function derived in the previous chapters we prove long time existence of vortex pairs. This result is optimal, since three vortices can collide in finite time. We report from our work with B. Gustafsson [52].

## 15.1 Planar hydrodynamics

We suppose that the flow region $\Omega \subset \mathbb{R}^2$ is bounded and finitely connected with Lipschitz boundary

$$\partial\Omega \;=\; \bigcup_{k=1}^{K} \Gamma_k .$$

We also suppose that it satisfies the uniform exterior ball condition. The velocity field at time $t$ is denoted by $v(t)$. First we investigate the time evolution of the scalar vorticity

$$\omega \;:=\; \operatorname{curl} v \;:=\; \partial_1 v^2 - \partial_2 v^1$$

in the smooth case. The velocity field of an incompressible fluid in two dimensions admits a stream function

$$v \;=\; J\nabla\psi$$

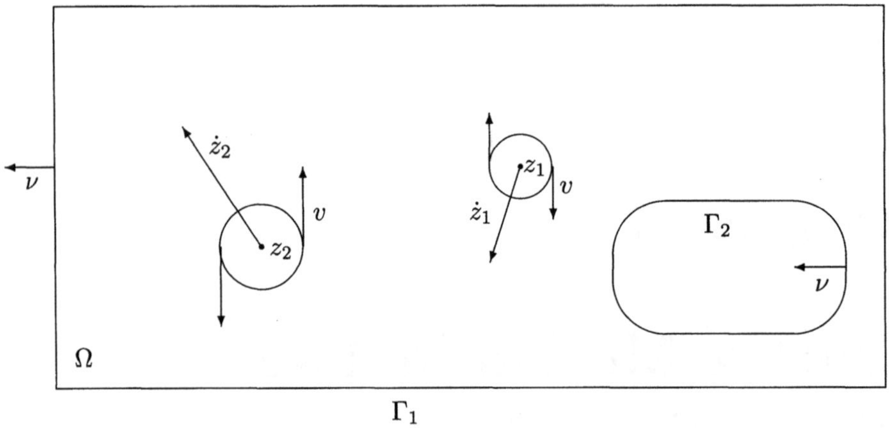


Figure 15.1: Flow region and point vortices.

where $J$ denotes the standard symplectic matrix. By definition of the vorticity the stream function is a solution of the Poisson equation

$$\begin{aligned} -\Delta\psi &= \omega \text{ in } \Omega, \\ \partial_{J\nu}\psi &= 0 \text{ on } \partial\Omega \end{aligned}$$

where $\partial_{J\nu}$ denotes the tangential derivative. The basic equation of non-viscous fluid motion are the Euler equations of ideal fluid dynamics

$$\partial_t v + (v \cdot \nabla)\, v + \frac{\nabla p}{\rho} = 0.$$

The density $\rho$ is constant. The convection term can be written as

$$(v \cdot \nabla)\, v = \frac{1}{2}\nabla |v|^2 - \omega J v.$$

In the stationary irrotational case we obtain Bernoulli's law

$$\frac{1}{2}|v|^2 + \frac{p}{\rho} = \text{const.}$$

Taking another curl of the Euler equations yields the transport equation for the vorticity. Using the incompressibility of the fluid one finds that the material derivative vanishes

$$D_t\omega := \partial_t\omega + (v \cdot \nabla)\, \omega = 0.$$

This means that the vorticity is convected along the flow. Moreover the *circulation*

$$c_k := \int_{\Gamma_k} v \cdot dz = \int_{\Gamma_k} \partial_\nu \psi$$

around each of the boundary components is conserved because

$$\dot{c}_k = \int_{\Gamma_k} \partial_t v \cdot dz = -\int_{\Gamma_k} \nabla\left(\frac{1}{2}|v|^2 + \frac{p}{\rho}\right) \cdot dz + \int_{\Gamma_k} \omega J v \cdot dz = 0.$$

This is Kelvin's theorem. Now we consider an initial vorticity distribution concentrated at finitely many points. It is represented by a collection of Dirac masses

$$\omega = \sum_{p=1}^{P} \omega_p \delta_{z_p}, \quad z_p \in \Omega.$$

The strength of the p'th vortex is $\omega_p$ and may be positive or negative according to its orientation (Figure 15.1). Strictly speaking the product term $\omega J v$ is not defined for such vorticity distributions. Even the weak form of the Euler equations is meaningless. Nevertheless $\omega J v$ vanishes except at the vortex centers. Thus the Euler equations simplify to

$$\partial_t v = -\nabla\left(\frac{1}{2}|v|^2 + \frac{p}{\rho}\right) \quad \text{in } \Omega \setminus \bigcup_p \{z_p(t)\}.$$

As a limit form of the equimeasurability property for smooth vortex distributions we postulate that the vorticity at any time $t \geq 0$ remains of the form

$$\omega(t) = \sum_p \omega_p \delta_{z_p(t)}.$$

## 15.2 Hydrodynamic Green's and Robin function

The properties of the stream function in the presence of point vortices suggests the definition of a special Green's function. It corresponds to the stream function of a single point vortex of unit strength.

**Definition 15.1 (Hydrodynamic Green's function)** *The hydrodynamic Green's function with periods $\gamma_1, \ldots, \gamma_K$ subject to $\sum \gamma_k = -1$ is defined as the solution $G_z$ of the problem*

$$\begin{aligned} -\Delta G_z &= \delta_z \quad \textit{in } \Omega, \\ \partial_{J\nu} G_z &= 0 \quad \textit{on } \partial\Omega, \\ \int_{\Gamma_k} \partial_\nu G_z &= \gamma_k \quad \textit{for every } k, \\ \int_{\partial\Omega} G_z \partial_\nu G_\zeta &= 0 \quad \textit{for every } z, \zeta \in \Omega. \end{aligned}$$

The hydrodynamic Green's function is constant on each boundary component. Its values will be denoted by $g_z^k := G_z|_{\Gamma_k}$. The last requirement is a normalization condition. It selects a unique solution (Lemma 15.3). The singularity at

$z$ represents a source of unit strength. Therefore necessarily $\sum \gamma_k = -1$. The first, second and third requirement specify $G_z$ up to an additive constant. The periods are independent of the normalization. Furthermore the hydrodynamic Green's function is symmetric. Integration by parts yields

$$G_z(\zeta) = \int_\Omega G_z(-\Delta G_\zeta) = -\int_\Omega \nabla G_z \cdot \nabla G_\zeta + \int_{\partial\Omega} G_z \partial_\nu G_\zeta$$

which is symmetric by the normalization. Uniqueness follows from the maximum principle. The regular part of the hydrodynamic Green's function is harmonic with the boundary values

$$\begin{aligned} H_z &= K_z - g_z^k \text{ on } \Gamma_k, \\ \int_{\Gamma_k} \partial_\nu H_z &= \int_{\Gamma_k} \partial_\nu K_z - \gamma_k. \end{aligned}$$

**Definition 15.2 (Hydrodynamic Robin function)** *The value of the regular part of the hydrodynamic Green's function at the singularity*

$$\tau(z) := H_z(z)$$

*is called hydrodynamic Robin function.*

The hydrodynamic Robin function is thus associated to the Laplacian acting on the class of functions which are locally constant on the boundary with periods $\gamma_k$. The difference of the hydrodynamic Green's function to the Dirichlet Green's function $G_z^0$ can be written in terms of capacity potentials or harmonic measures $u_k$ defined by

$$\begin{aligned} \Delta u_k &= 0 \text{ in } \Omega, \\ u_k &= \delta_{kl} \text{ on } \Gamma_l. \end{aligned}$$

These functions form a basis of the space of harmonic functions which are locally constant on the boundary. They add up to unity. By the general hypotheses on the domain each $u_k$ has bounded gradient.

**Lemma 15.3 (Properties of hydrodynamic Green's function)** *Under the general hypotheses of Section* 15.1 *on* $\Omega$ *we have:*

1. *The hydrodynamic Green's function exists if and only if the periods satisfy the consistency relation*

(15.1)
$$\sum_k \gamma_k = -1.$$

2. *The hydrodynamic Green's function is unique, symmetric and*

(15.2)
$$G_z(\zeta) = G_z^0(\zeta) + \sum_{k,l} g^{kl} u_k(z) u_l(\zeta),$$

*where* $g^{kl}$ *is a symmetric positive semi definite matrix with one dimensional kernel spanned by the vector* $(\gamma_1, \dots, \gamma_K)$.

3. *At the boundary the hydrodynamic Robin function blows up like*

$$\tau(z) \quad = \quad -\frac{1}{2\pi}\log(d(z)) + O(1)$$

*uniformly as $z$ tends to $\partial\Omega$.*

*Proof.* Symmetry and uniqueness are already proved. The necessity of (15.1) follows from

$$0 \quad = \quad \int_{\Omega\setminus B_z^r} \Delta G_z \quad = \quad \sum_k \int_{\Gamma_k} \partial_\nu G_z - \int_{\partial B_z^r} \partial_\nu G_z \quad = \quad 1 + \sum_k \gamma_k$$

for $r > 0$ sufficiently small. The difference between the two Green's functions

$$G_z - G_z^0 \quad = \quad H_z^0 - H_z$$

is harmonic and locally constant on the boundary. The values of the hydrodynamic Green's function on the respective boundary components are $g_z^k$. Thus

$$G_z - G_z^0 \quad = \quad \sum_k g_z^k u_k.$$

By symmetry of both Green's functions

$$g_z^k \quad = \quad \sum_l g^{kl} u_l(z)$$

with a symmetric matrix $g^{kl}$. To construct $G_z$ we take (15.2) as an ansatz and try to determine the matrix $g^{kl}$. This will prove existence. The first two properties of $G_z$ in Definition 15.1 are automatically satisfied. For the remaining properties we need the following facts. The capacity coefficients

$$p_{kl} \quad = \quad \int_{\Gamma_k} \partial_\nu u_l \quad = \quad \int_{\partial\Omega} u_k \partial_\nu u_l \quad = \quad \int_\Omega \nabla u_k \cdot \nabla u_l \tag{15.3}$$

form a symmetric, positive semi definite matrix with one-dimensional kernel

$$\ker(p) \quad = \quad \operatorname{span}(1, \ldots, 1)$$

because $\sum_k u_k = 1$. Moreover

$$u_k(z) \quad = \quad -\int_{\partial\Omega} u_k \partial_\nu G_z^0 \quad = \quad -\int_{\Gamma_k} \partial_\nu G_z^0. \tag{15.4}$$

Using (15.3) and (15.4) the circulation property of $G_z$ reduces to

$$\begin{aligned} \gamma_k \quad &= \quad \int_{\Gamma_k} \partial_\nu G_z \quad = \quad \int_{\Gamma_k} \partial_\nu G_z^0 + \sum_{i,j} g^{ij} u_i(z) \int_{\Gamma_k} \partial_\nu u_j \\ &= \quad -u_k(z) + \sum_{i,j} p_{kj} g^{ij} u_i(z) \quad = \quad \sum_i \left( \sum_j p_{kj} g^{ij} - \delta_{ki} \right) u_i(z). \end{aligned}$$

By linear independence of the $u_i$'s and $\sum_i u_i = 1$ the term in bracket must be $\gamma_k$. Therefore

$$\sum_j p_{kj} g^{ij} = \gamma_k + \delta_{ki} \tag{15.5}$$

for every $i$ and $k$. For fixed $i$ this is a linear equation for the vector $\left(g^{ij}\right)_{j=1,\ldots,K}$. It is uniquely solvable if and only if the vector on the right is orthogonal to the kernel of the capacity matrix $(p_{kj})$

$$\sum_k 1 \cdot (\gamma_k + \delta_{ki}) = 0,$$

i.e. if (15.1) holds. Moreover every solution is of the form $\left(g^{ij} + c_i\right)_{j=1,\ldots,K}$. We can adjust the constant such that

$$\sum_j \gamma_j g^{ij} = 0$$

for each $i$. This determines $\left(g^{ij}\right)$ uniquely. The above equation is the normalization condition. The symmetry of $\left(g^{ij}\right)$ follows from (15.5) and symmetry of $(p_{kl})$:

$$g^{ji} = \sum_k (\gamma_k + \delta_{ki})\, g^{jk} = \sum_{k,l} p_{kl} g^{jk} g^{il}.$$

Concerning the definiteness consider the quadratic form

$$Q(\xi) := \sum_{i,j} g^{ij} \xi_i \xi_j = \sum_{k,l} p_{kl} \left(\sum_j g^{jk} \xi_j\right) \left(\sum_i g^{il} \xi_i\right).$$

By the properties of $(p_{kl})$ this form is positive and $Q(\xi) = 0$ if and only if $\sum_j g^{jk} \xi_j$ is a constant independent of $k$. When this occurs (15.5) gives

$$0 = \sum_{i,l} p_{kl} g^{il} \xi_i = \sum_i (\gamma_k + \delta_{ki})\, \xi_i = \xi_k + \gamma_k \sum_i \xi_i,$$

i.e. $\xi_k = c\gamma_k$. This means that the kernel of $\left(g^{il}\right)$ is spanned by $(\gamma_1, \ldots, \gamma_K)$. In particular $\left|G_z - G_z^0\right| \leq \|g\|$ and therefore $\tau^0 - \|g\| \leq \tau \leq \tau^0$ where

$$\|g\| := \sup \left\{ \sum_{k,l} g^{kl} \xi_k \xi_l : \sum_k |\xi_k| \leq 1 \right\}$$

which is finite by finite connectivity of $\Omega$. Thus the boundary behaviour of the two Robin functions is the same up to a bounded difference and the third assertion follows from Lemma 8.3. □

## 15.3 Point vortex model

Now we are in position to formulate the equation of motion for the vortex centers. For $\Omega = \mathbb{R}^2$ it goes back to the work of Kirchhoff in 1876. The self interaction term in the presence of solid boundaries has been discovered by Routh in 1881. The renormalized kinetic energy is also called *Kirchhoff-Routh path function.* In 1943 C.C. Lin derived the equation of motion in the most general case.

**Theorem 15.4 (Kirchhoff [75], Routh [115], Lin [87])** *Let $\Omega$ be a domain satisfying the general hypotheses of Section* 15.1 *and suppose that the initial vorticity distribution is of the form $\omega(0) = \sum_p \omega_p \delta_{z_p(0)}$ with total vorticity $\sum_p \omega_p \neq 0$. Given the circulations $c_k$ satisfying the consistency relation $\sum_k c_k + \sum_p \omega_p = 0$ define the periods*

$$\gamma_k \quad := \quad \frac{c_k}{\sum_p \omega_p}$$

*and the corresponding hydrodynamic Green's function $G_z$ and Robin function $\tau$ as in the Definitions* 15.1 *and* 15.2. *Then:*

1. *The speed of the vortex centers is given by*

$$\dot{z}_p \quad = \quad \sum_{q \neq p} \omega_q J \nabla G_{z_q}(z_p) - \frac{\omega_p}{2} J \nabla \tau(z_p). \tag{15.6}$$

2. *The renormalized kinetic energy*

$$E(z_1, \ldots, z_P) \quad := \quad \sum_{\{p,q\,:\,q>p\}} \omega_p \omega_q G_{z_q}(z_p) - \frac{1}{2} \sum_p \omega_p \tau(z_p)$$

   *is an integral of motion.*

3. *The equation of motion* (15.6) *has a Hamiltonian structure with Hamiltonian $E$ and symplectic form $d\Omega = \sum_p \omega_p dx_p dy_p$. I.e.*

$$\dot{z}_p \quad = \quad \frac{1}{\omega_p} J \nabla_{z_p} E.$$

*Proof.* The stream function is given by

$$\psi \quad = \quad \sum_p \omega_p G_{z_p}.$$

By construction it has the prescribed circulations $c_k$. Regarding the velocity field we single out the contribution of the p'th vortex and decompose it into singular and regular part

$$v \quad = \quad \omega_p J \nabla K_{z_p} - \omega_p j \nabla H_{z_p} + \sum_{q \neq p} \omega_q j \nabla G_{z_q}.$$

The term $J\nabla K_{z_p}$ describes a pure rotation around the center $z_p$ which does not contribute to the motion of $z_p$ itself. Only the regular part contributes to its drift

$$\dot{z}_p = v_{\text{reg}}(z_p) = \sum_{q \neq p} \omega_q j \nabla G_{z_q}(z_p) - \omega_p j \nabla H_{z_p}(z_p).$$

The equation of motion (15.6) follows from the relation $\nabla \tau(z) = 2\nabla H_z(z)$ which holds by symmetry of the hydrodynamic Green's function. Conservation of $E$ is a consequence of the equation of motion and antisymmetry of the symplectic form. □

The argument can be considered heuristic. On the other hand it is easy to verify conservation of the renormalized kinetic energy. However, this calculation does not explain how the expression for the renormalized kinetic energy can be found. In Section 15.4 we show how the core energy method can be used for this purpose.

**Example 15.5** For $\Omega = \mathbb{R}^2$ Theorem 15.4 does not apply directly. Nevertheless the conclusions hold under rest conditions at infinity with $G_z = K_z$ and $\tau = 0$. The equation of motion for two vortices in the plane is

$$\dot{z}_1 = \frac{\omega_2}{2\pi} J \frac{z_2 - z_1}{|z_2 - z_1|^2}, \quad \dot{z}_2 = \frac{\omega_1}{2\pi} J \frac{z_1 - z_2}{|z_1 - z_2|^2}, \quad E = -\frac{\omega_1 \omega_2}{2\pi} \log |z_1 - z_2| .$$

In particular the distance of the vortex centers remains constant and the center of mass $\omega_1 z_1 + \omega_2 z_2$ is at rest. The motion of an arbitrary number of point vortices in the plane admits three independent integrals which are in involution with respect to the symplectic form $d\Omega$, namely

$$E, \left| \sum_p \omega_p z_p \right|^2, \sum_p \omega_p |z_p|^2 .$$

The corresponding conserved quantities (symmetries) are energy (time shift), moment of inertia (translation) and angular momentum (rotation). Thus the motion of up to three vortices is integrable while the motion of 4 and more vortices may be chaotic depending on the initial data (Aref and Pomphrey [9], Aref [8]).

**Example 15.6** For the upper half plane $\{z = x + iy \,:\, y > 0\}$ one has

$$G_z(\zeta) = -\frac{1}{2\pi} \log \left| \frac{\zeta - z}{\zeta - \bar{z}} \right|, \quad \tau(z) = -\frac{1}{2\pi} \log (2y) .$$

A single vortex moves parallel to the boundary. Also the motion of two vortices of opposite orientation at equal distance from the boundary $z_{1,2} = \pm x + iy$ can be computed easily. We suppose no flux conditions at infinity. The Robin

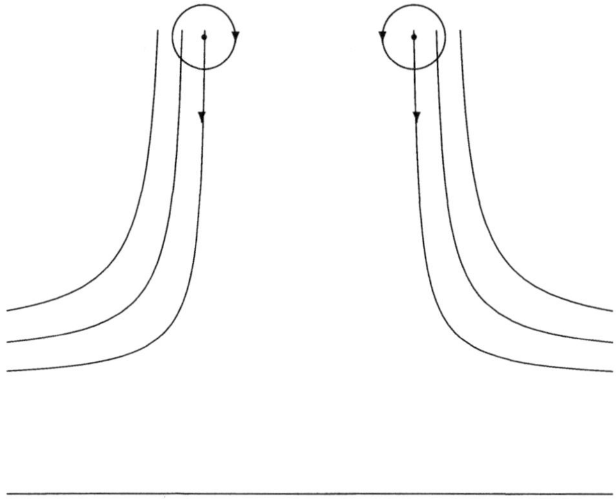

Figure 15.2: Motion of two vortices of opposite orientation in a half plane.

function of the quarter plane $\{x+iy : x>0,\ y>0\}$ is $\tau(z) = -\frac{1}{4\pi}\log\frac{2x^2y^2}{x^2+y^2}$. Reflection at the y-axis gives the renormalized energy for the half plane

$$E(z_1, z_2) = \frac{\omega^2}{4\pi}\log\frac{2x^2y^2}{x^2+y^2}$$

The vortex trajectories are of the form $\left\{x^2y^2 = c^2\left(x^2+y^2\right)\right\}$ (Figure 15.2). The parameter $c$ gives the asymptotic distance from both axes.

## 15.4 Core energy method

The law of motion for the vortex centers can be found by the core energy method or energy renormalization. The idea is to separate the energy into two parts each of which is conserved separately. The infinite part (the core energy) is the same for all geometries, while the remaining part (the renormalized energy) is the interesting one that carries geometrical information. The core energy method is a heuristic way of deriving a finite conserved quantity from an infinite energy. The guess obtained in this way can be verified a posteriori using the equations of motion. An alternate derivation based on the calculus of variations is given in [52]. For simplicity we restrict ourselves to the case of a single vortex of unit strength, i.e. $\psi = G_z$. The kinetic energy of the flow is

$$E_{\text{kin}} = \frac{1}{2}\int_\Omega |v|^2 = \frac{1}{2}\int_\Omega |\nabla G_z|^2 = \infty.$$

The unbounded contribution comes from the core of the vortex. This suggests the excision of a small ball $B_z^r$ around the center, dividing the kinetic energy into a finite and an infinite part. Since the fluid is incompressible and circulates around the center only a small amount of kinetic energy can cross the artificial boundary $\partial B_z^r$ in finite time. This means that the energy diffusion between the finite and the infinite part of the energy can be neglected in the limit as the radius of the ball tends to zero. The finite part of the energy is

$$\begin{aligned}
\int_{\Omega\setminus B_z^r} |\nabla G_z|^2 &= \int_{\partial(\Omega\setminus B_z^r)} G_z\partial_\nu G_z = -\int_{\partial B_z^r} G_z\partial_\nu G_z \\
&= -\int_{\partial b_z^r} K_z\partial_\nu K_z + \int_{\partial b_z^r} K_z\partial_\nu H_z \\
&\quad + \int_{\partial b_z^r} H_z\partial_\nu K_z - \int_{\partial b_z^r} H_z\partial_\nu H_z \\
&= -\frac{1}{2\pi}\log(r) + 0 - \tau(z) - \int_{B_z^r} |\nabla H_z|^2 \\
&= -\frac{1}{2\pi}\log(r) - \tau(z) + O\left(r^2\right).
\end{aligned}$$

The gradient of this expression is

$$\begin{aligned}
\nabla_z \int_{\Omega\setminus B_z^r} |\nabla G_z|^2 &= -\nabla\tau(z) - \nabla_z \int_{B_z^r} |\nabla H_z|^2 \\
&= -\nabla\tau(z) - \int_{\partial b_z^r} |\nabla H_z|^2 \nu - \int_{B_z^r} \nabla_z |\nabla H_z|^2 \\
&= -\nabla\tau(z) + O(r).
\end{aligned}$$

Thus the time derivative of $\int_{\Omega\setminus B_z^r} |\nabla G_z|^2$ is equal to that of $\tau(z)$ up to an error term which tends to zero as $r \to 0$. We conclude that $E = -\frac{1}{2}\tau$ is conserved along the flow. In the case of multiple vortices the only difference is an additional contribution to the regular part due to the other vortices obtained from the relation $\int_\Omega \nabla G_z \cdot \nabla G_\zeta = G_z(\zeta)$.

The motion of point vortices can be seen as the limit of several more realistic moving boundary problems [52]. Each of them justifies the core energy method as a method for the derivation of the point vortex model.

## 15.5 Motion of isolated point vortices

In this section we discuss qualitative properties of the point vortex model that can be derived from conservation of the renormalized kinetic energy. We adopt the general hypotheses on the domain of Section 15.1.

**Corollary 15.7 (Motion of a single vortex)** *The center $z$ of a single vortex of strength $\omega$ behaves as follows.*

1. *It moves along the level lines of the hydrodynamic Robin function at speed*

$$\dot{z} = -\frac{\omega}{2}J\nabla\tau(z).$$

*In particular almost all orbits are periodic.*

2. *Every critical point of the hydrodynamic Robin function is a rest point. There is at least one.*

3. *A vortex center close to a local minimum $z_0$ of the hydrodynamic Robin function circulates along the boundary of a small "ellipse". As the ellipse shrinks to a point the time of revolution tends to*

$$T = \frac{4\pi}{\omega\sqrt{\det D^2\tau(z_0)}}.$$

*Proof.* The first and second claim are immediate consequences of Theorem 15.4 and Lemma 15.3. For the third one we normalize: $z_0 = 0$, $\tau(z_0) = 0$ and $\tau(z) = \frac{1}{2}\left(\tau_{11}z_1^2 + \tau_{22}z_2^2\right) + O(|z|^3)$ as $|z| \to 0$. For small $h > 0$ the area of the "ellipse" is

$$|\{t < h\}| = \frac{2\pi h}{\sqrt{\tau_{11}\tau_{22}}} + O\left(h^2\right).$$

By the co-area formula the corresponding time of revolution is

$$\int_{\partial\{t<h\}} \frac{ds}{|\dot{z}|} = \frac{2}{\omega}\int_{\partial\{t<h\}} \frac{ds}{|\nabla t|} = \frac{2}{\omega}\partial_h\,|\{t < h\}|.$$

□

Figure 8.3 shows the trajectories of a single vortex.

**Corollary 15.8 (Motion of multiple vortices)** *The following types of collisions are excluded as long as no other collisions occur simultaneously.*

1. *Collision of two vortices in the interior of the domain.*

2. *Collision of a single vortex with the boundary.*

3. *Collision of multiple vortices in the interior of the domain unless*

$$\sum_{q\neq p} \omega_p\omega_q \log|z_p - z_q|$$

*has a finite limit.*

4. *The motion of two vortices of different orientation exists for all time and admits a stationary constellation at* $\max E(z_1, z_2)$.

5. *If the total vorticity $\omega := \sum \omega_p$ does not vanish then the speed of the center of vorticity*

$$z := \frac{1}{\omega}\sum \omega_p z_p$$

*is given by*

$$\dot{z} = -\frac{\omega}{2}\int_{\partial\Omega} |v|^2\, J\nu.$$

*Proof.* All claims follow from conservation of the renormalized kinetic energy and the boundary behaviour of the hydrodynamic Robin function; the last one from Lemma 8.4. □

At a collision the vortex model breaks down. The vector field is no longer Lipschitz. In particular there is no unique continuation. An example of a triple collision in $\mathbb{R}^2$ can be found in Kimura [74] and in the book of Marchioro and Pulvirenti [93]. In this example the relative distances of the vortices are preserved and $\omega_1\omega_2 + \omega_2\omega_3 + \omega_3\omega_1 = 0$. Thus the singular contributions to the renormalized energy cancel at the collision. The collision happens in finite time and the velocities of the vortex centers are unbounded. Much more can be said in the case of a single vortex moving in a simply connected domain. In this case the normalization yields $G_z = 0$ on $\partial\Omega$, i.e. $G_z = G_z^0$ and $\tau = \tau^0$. In particular a vortex center at a small distance $d$ from the boundary moves with speed

$$\dot{z} \;=\; -\frac{\omega}{4\pi d} J\nu + O(1) \;\text{ as }\; d \to 0$$

as follows from Lemma 8.6. The same lemma provides a minimal distance of a stationary vortex from the walls.

## 15.6 Motion of vortex clusters

First we show that on a macroscopic scale the motion of a vortex cluster converges to that of a point vortex as the cluster shrinks to a point. Consider a cluster $\{z_p : p \in C\}$ of non-vanishing total vorticity $\omega := \sum_{p\in C} \omega_p$ with center of vorticity

$$z \;:=\; \frac{1}{\omega} \sum_{p\in C} \omega_p z_p.$$

Let $r$ denote the diameter of the cluster. Obviously $z_p \in B_z^{\theta r}$ for a suitable factor $\theta$. We show that the center of vorticity essentially moves like a single vortex of vorticity $\omega$. The dependence of the error term on the distance from the boundary is essential for our stability result for vortex pairs.

**Theorem 15.9 (Macroscopic cluster dynamics)** *Assume the general hypotheses of Section* 15.1 *on* $\Omega$ *and that the vortex cluster* $\{z_p : p \in C\}$ *of non-vanishing total vorticity has small diameter and is well separated from the other vortices, i.e.* $|z_p - z_{p'}| < r$ *for* $p, p' \in C$ *and* $|z_p - z_q| > R$ *for* $p \in C$, $q \notin C$ *where* $r \ll R$ *and* $R$ *is kept fixed. Then the center of vorticity moves according to*

$$\dot{z} \;=\; \sum_{p\notin C} \omega_p J\nabla G_{z_p}(z) - \frac{\omega}{2} J\nabla\tau(z) + O\left(\frac{r^2}{d(z)^3}\right) \quad \textit{as} \;\; r \to 0.$$

*Proof.* By the equation of motion (15.6) the center of vorticity moves according to

$$
\begin{aligned}
\omega \dot{z} &= \sum_{p\in C} \omega_p \dot{z}_p \\
&= \sum_{p\in C}\sum_{q\neq p} \omega_p\omega_q J\nabla G_{z_q}(z_p) - \frac{1}{2}\sum_{p\in C}\omega_p^2 J\nabla\tau(z_p) \\
&= \sum_{p\in C}\sum_{q\notin C} \omega_p\omega_q J\nabla G_{z_q}(z_p) - \sum_{p,p'\in C}\omega_p\omega_{p'} J\nabla H_{z_{p'}}(z_p) \qquad (15.7)
\end{aligned}
$$

because $\nabla K_{z_q}(z_p) + \nabla K_{z_p}(z_q) = 0$ and $\nabla\tau(z) = 2\nabla H_z(z)$. On the other hand the right side in Theorem 15.9 multiplied with $\omega$ is

$$
\sum_{p\in C}\sum_{q\notin C} \omega_p\omega_q J\nabla G_{z_q}(z) - \sum_{p,p'\in C}\omega_p\omega_{p'} J\nabla H_z(z). \qquad (15.8)
$$

The difference between (15.7) and (15.8) is

$$
\begin{aligned}
&\sum_{p\in C}\sum_{q\notin C} \omega_p\omega_q J\left(\nabla G_{z_q}(z_p) - \nabla G_{z_q}(z)\right) \\
&- \sum_{p,p'\in C}\omega_p\omega_{p'} J\left(\nabla H_{z_{p'}}(z_p) - \nabla H_z(z)\right).
\end{aligned}
$$

The first term is

$$
\sum_{q\notin C}\omega_q\sum_{p\in C}\omega_p J D^2 G_{z_q}(z)(z_p - z) + O\left(r^2\right) \;=\; O\left(r^2\right)
$$

because $\sum_{p\in C}\omega_p(z_p - z) = 0$. The second term satisfies the same estimate. More precisely the total error is of order

$$
r^2 \sup_{\zeta\in B_z^{\theta r}}\left(\sup_{q\notin C}\|\nabla D^3 G_{z_q}(\zeta)\| + \sup_{p\in C}\|\nabla D^3 H_z(z_p)\| + \|\nabla D^3 H_z(\zeta)\|\right).
$$

By Lemma 15.10 below this is of order $O\left(r^2/d(z)^3\right)$. □

Theorem 15.9 does not apply to Example 15.6 where $\omega = 0$. In this case the motion is governed by higher order derivatives of the Robin function.

**Lemma 15.10** *If $\Omega$ satisfies the hypotheses of Section* 15.1 *then*

$$
\left|\nabla_z^m\nabla_\zeta^n H_z(\zeta)\right| \;\le\; \frac{c_\Omega(m,n)}{d(z)^m d(\zeta)^n}
$$

*for every $m+n\geq 1$.*

*Proof.* Recall that the regular part is symmetric and harmonic in each variable. At the boundary

$$H_z \quad = \quad K_z - \sum_j g^{kj} u_j(z) \quad \text{on} \quad \Gamma_k.$$

Thus

$$H_z(\zeta) \quad = \quad -\sum_k \int_{\Gamma_k} \left( K_z - \sum_j g^{kj} u_j(z) \right) \partial_\nu G^0_\zeta.$$

Using

$$|\nabla_z K_z(w)| \quad = \quad \frac{1}{2\pi |w-z|} \quad \leq \quad \frac{1}{2\pi d(z)} \quad \text{for} \quad w \in \partial\Omega$$

together with $|\nabla u_j| \leq c$, $-\partial_\nu G^0_\zeta \geq 0$ and $-\int_{\partial\Omega} \partial_\nu G^0_\zeta = 1$ we get

$$|\nabla_z H_\zeta(z)| \quad = \quad |\nabla_z H_z(\zeta)| \quad \leq \quad \frac{c}{d(z)} \quad \text{for every} \quad \zeta \in \partial\Omega.$$

By the maximum principle this yields the desired estimate for $m+n=1$. If $h$ is harmonic in $B^r_z$ then

$$|\nabla h(z)| \quad \leq \quad \frac{1}{r} \sup_{\partial B^r_z} |h|\,.$$

Application of this estimate to $h = \nabla H_\zeta$ on $B^r_z \subset B^{2r}_z \subset \Omega$ proves the lemma for $m+n=2$. Similarly we estimate higher order derivatives. □

Now we can analyze the dynamics within a vortex cluster normalized with respect to the center of vorticity.

**Theorem 15.11 (Microscopic cluster dynamics)** *Under the hypotheses of Theorem* 15.9 *and* $d(z) \geq r > 0$ *we have*

$$(z_p - z)^{\cdot} \quad = \quad \sum_{p' \in C,\ p' \neq p} \omega_{p'} J \nabla K_{z_{p'}}(z_p) + O\left(\frac{r}{d(z)^2}\right) \quad \text{as} \quad r \to 0$$

*for every* $p \in C$.

*Proof.* By (15.6), Theorem 15.9 and Lemma 15.10 we have

$$\begin{aligned}(z_p - z)^{\cdot} \quad = \quad & \sum_{q \notin C} \omega_q J \left(\nabla G_{z_q}(z_p) - \nabla G_{z_q}(z)\right) + \sum_{p' \in C,\, p' \neq p} \omega_{p'} J \nabla G_{z_{p'}}(z_p) \\ & - \frac{\omega_p}{2} J \nabla \tau(z_p) + \frac{\omega}{2} J \nabla \tau(z) + O\left(\frac{r^2}{d(z)^3}\right)\end{aligned}$$

$$
\begin{aligned}
&= \sum_{p' \in C, p' \neq p} \omega_{p'} J \nabla K_{z_{p'}}(z_p) \\
&\quad - \sum_{p' \in C} \omega_{p'} J \left( \nabla H_{z_{p'}}(z_p) - \nabla H_z(z) \right) + O\left( \frac{r}{d(z)^2} \right) \\
&= \sum_{p' \in C,\ p' \neq p} \omega_{p'} J \nabla K_{z_{p'}}(z_p) + O\left( \frac{r}{d(z)^2} \right).
\end{aligned}
$$

□

Also this formula is second order accurate because the leading term is of order $O(1/r)$.

## 15.7 Stability of vortex pairs

We show that two vortices do not collide with each other nor with the boundary and that they stay close together if they are close initially. In particular there is no recombination of vortex pairs of opposite strength. If further vortices are present the same holds as long as no other collisions occur. The difficult step in this proof is exclusion of collisions of equally oriented vortices at the boundary. As an illustration we first consider the motion of two equally oriented vortices $z_1 = (x_1, y_1)$ and $z_2 = (x_2, y_2)$ in the half-space $\Omega = \{z \in \mathbb{R}^2 : y > 0\}$. The Hamiltonian is translation invariant in the $x$-direction, i.e.

$$
0 = \frac{d}{dx} E\left( z_1 + \begin{pmatrix} x \\ 0 \end{pmatrix}, z_2 + \begin{pmatrix} x \\ 0 \end{pmatrix} \right)\Bigg|_{x=0} = (\partial_{x_1} + \partial_{x_2}) E.
$$

Thus the linear momentum $\omega_1 y_1 + \omega_2 y_2$ is conserved:

$$
(\omega_1 y_1 + \omega_2 y_2)^{\cdot} = -\partial_{x_1} E - \partial_{x_2} E = 0.
$$

This is an instant of Noether's theorem. Since $\omega_1 \omega_2 > 0$ the coordinates $y_1$ and $y_2$ cannot go to zero simultaneously.

**Theorem 15.12 ([52])** *The motion of two vortices in a domain satisfying the uniform exterior ball condition exists for all time. In particular they do not collide in finite time. In all cases $|z_1(t) - z_2(t)| \leq D\,|z_1(0) - z_2(0)|$ with a constant $D$ depending only on $\Omega$.*

1. *If $\omega_1 \omega_2 < 0$ then the motion stays away from the boundary of the phase space*

$$
\{(z_1, z_2) \in \Omega \times \Omega : z_1 \neq z_2\}
$$

*by a uniform distance.*

2. *If $\omega_1\omega_2 > 0$ and one of the vortices tends to the boundary then so does the other and*

$$|z_1 - z_2| \;\leq\; C\, d(z_p)^{2+\alpha},$$

$$d\left(z_p(t)\right) \;\geq\; \begin{cases} Ae^{-Bt} & (always), \\ (A+Bt)^{-\frac{1}{2\alpha}} & (\omega_1 \neq \omega_2) \end{cases}$$

*where*

$$\alpha \;:=\; \frac{\left(\omega_1 - \omega_2\right)^2}{2\omega_1\omega_2} \;\geq\; 0.$$

*for $p = 1, 2$ and $t \geq 0$ with positive constants $A$, $B$, $C$.*

*Proof.* If $\omega_1\omega_2 < 0$ then each term in

$$E(z_1, z_2) \;=\; \omega_1\omega_2 G_{z_1}(z_2) - \frac{1}{2}\,\omega_1^2 t\,(z_1) - \frac{1}{2}\,\omega_2^2 \tau(z_2)$$

is bounded above. By conservation of $E$ they are also bounded below. From the asymptotics for the Robin function at the boundary (Lemma 15.3) we get a lower bound $d(z_p) \geq \delta > 0$. On the remaining set boundedness of $G_{z_1}(z_2)$ leads to a uniform lower bound for $|z_1 - z_2|$. Energy conservation also implies that

$$-\log|z_1(t) - z_2(t)| \;=\; -\log|z_1(0) - z_2(0)| + g(t)$$

with a bounded function $g$ whose bounds only depend on $\Omega$.

For equally oriented vortices energy conservation does not lead to a uniform distance from the boundary of the phase space. However, we already know that the vortices do not collide in the interior and that none of them tends to the boundary alone. Thus we can restrict our attention to the case that they tend to the boundary simultaneously. We use Theorem 15.9 and adopt the notation introduced thereby. Note that the total vorticity is nonzero. By energy conservation and Lemma 15.3 we have

$$G^0_{z_1}(z_2) - \frac{\omega_1}{2\omega_2}\tau^0\,(z_1) - \frac{\omega_2}{2\omega_1}\tau^0(z_2) \;\geq\; \frac{E}{\omega_1\omega_2} - c. \tag{15.9}$$

Comparison with the Dirichlet Green's function of the complement of a disk $(B^r_\zeta)^c \supset \Omega$ yields

$$-\frac{1}{2\pi}\log\left|\frac{r\,(z_1 - z_2)}{r^2 - (z_2 - \zeta)\,\overline{(z_1 - \phantom{\zeta} z)}}\right| \;\geq\; G^0_{z_1}(z_2).$$

Choosing $\zeta$ such that $|z_2 - \zeta| = d(z_2) + r$ we can estimate

$$\begin{aligned} \left|r^2 - (z_2 - \zeta)\,\overline{(z_1 - \zeta)}\right| \;&\leq\; |z_2 - \zeta|^2 - r^2 + |(z_2 - \zeta)\,(\overline{z_1} - \overline{z_2})| \\ &\leq\; c\left(d(z_2) + |z_1 - z_2|\right). \end{aligned}$$

By Lemma 15.3 exponentiation of the energy inequality (15.9) yields

$$\frac{|z_1 - z_2|}{d(z_p) + |z_1 - z_2|} \quad \leq \quad C e^{-\frac{2\pi}{\omega_1 \omega_2} E} d\,(z_1)^{\frac{\omega_1}{2\omega_2}}\, d(z_2)^{\frac{\omega_2}{2\omega_1}}$$

for $p = 2$. By symmetry also for $p = 1$. In particular

$$\frac{|z_1 - z_2|}{d(z_p)} \quad \to \quad 0 \;\text{ as }\; z_1 \;\text{ or }\; z_2 \to \partial\Omega$$

and so

$$\frac{|z_1 - z_2|}{d(z_p)} \quad \leq \quad C e^{-\frac{2\pi}{\omega_1 \omega_2} E} d\,(z_1)^{\frac{\omega_1}{2\omega_2}}\, d(z_2)^{\frac{\omega_2}{2\omega_2}}.$$

By the exterior ball condition also $z \in \Omega$, $d(z) \to 0$ and we can replace $d(z_p)$ by $d(z)$ up to lower order terms. This yields

$$(15.10) \qquad |z_1 - z_2| \quad \leq \quad C d(z)^{2+\alpha} \;\text{ as }\; z_1 \;\text{ or }\; z_2 \to \partial\Omega$$

because $\frac{\omega_1}{2\omega_2} + \frac{\omega_2}{2\omega_1} + 1 = 2 + \alpha$. For convex domains the same estimate extends to all of $\Omega \times \Omega$. According to Theorem 15.9 the center of vorticity moves with speed

$$\dot{z} \quad = \quad -\frac{\omega}{2} J \nabla \tau(z) + O\left( \frac{|z_1 - z_2|^2}{d(z)^3} \right)$$

which is orthogonal to $\nabla\tau(z)$ up to an error which by (15.10) is of order $O(d(z)^{1+2\alpha})$. Therefore

$$\partial_t \tau(z(t)) \quad = \quad \nabla\tau(z(t)) \cdot \dot{z}(t) \quad \leq \quad C d(z)^{2\alpha}$$

by Lemma 15.10. Lemma 15.3 together with elementary comparison arguments for ordinary differential equations leads to the lower bound for $d(z)$. □

Thus a pair of equally oriented vortices approaching the boundary rotates rapidly around their common center of vorticity and slides along one of the boundary components at exponentially increasing speed.

## 15.8 Numerical approximation of vortex motion

On a simply connected domain the orbits of a single vortex center are easily obtained by solving Liouville's equation (Section 8.4.2). The level sets of $\tau$ are the desired trajectories (Figures 8.2 and 8.3).

For multiply connected domains and multiple vortices we use a variant of the boundary element method of Section 8.4.1. This approach is based on harmonicity of the regular part $H_z$ of the hydrodynamic Green's function whose value and gradient is needed at $P^2$ points. We use the single layer representation

$$\int_{\partial\Omega} q_z K_\zeta \;=\; \begin{cases} H_z(\zeta), & \zeta \in \overline{\Omega}, \\ H_z^c(\zeta), & \zeta \in \overline{\Omega^c} \end{cases}$$

with the harmonic function $H_z^c$ defined by this relation. The density $q_z$ represents the jump of the normal derivative

$$q_z \;=\; \partial_\nu H_z - \partial_\nu H_z^c$$

while $H_z = H_z^c$ on $\partial\Omega$. If $\int_{\partial\Omega} q_z = 0$ then $H_z^c$ is regular at infinity. This can be seen as follows. Let $h$ be the regular harmonic extension of $H_z|_{\partial\Omega}$ to $\Omega^c \cup \{\infty\}$. Then $g := H_z^c - h$ is harmonic in $\Omega^c$, vanishes at the boundary and $\lim_{\zeta\to\infty} g(\zeta)$ exists in $\overline{\mathbb{R}}$. Thus $g = \lambda G_\infty$ must be a constant multiple of the Dirichlet Green's function on $\Omega^c$ with singularity $-K_0$ at infinity. This implies $\lambda = 0$ because

$$0 \;=\; \int_{\partial\Omega} q_z \;=\; \int_{\partial\Omega} \partial_\nu H_z - \int_{\Gamma_0} \partial_\nu \left(\lambda G_\infty + h\right) \;=\; -\lambda$$

where $\Gamma_0$ denotes the exterior boundary component. In particular

$$\int_{\Gamma_k} q_z \;=\; \int_{\Gamma_k} \left(\partial_\nu H_z - \partial_\nu H_z^c\right) \;=\; \int_{\Gamma_k} \partial_\nu H_z$$

for every $k$. Using the decomposition $G_z = K_z - H_z$ and $\nabla_\zeta K_\zeta(z) = -\nabla_z K_\zeta(z)$ the definition of the hydrodynamic Green's function (Definition 15.1) translates into the subsequent scheme.

1. Discretize the boundary $\partial\Omega$. On every boundary component choose an arbitrary point $\zeta_k \in \Gamma_k$. Fix a time step $dt$.

2. Suppose we know the position of the vortex centers at time $t$. For $z = z_1, \ldots, z_P$ solve

$$\begin{aligned} \text{(15.11)} \quad \int_{\partial\Omega} q_z \partial_{J\nu(\zeta)} K_\zeta &= \partial_{J\nu(\zeta)} K_z(\zeta) \qquad (\zeta \in \partial\Omega), \\ \int_{\Gamma_k} q_z &= \int_{\Gamma_k} \partial_\nu K_z - \gamma_k \qquad (k = 1, \ldots, K), \\ \int_{\partial\Omega} q_z \sum_k \gamma_k K_{\zeta_k} &= \sum_k \gamma_k K_{\zeta_k}(z) \end{aligned}$$

for $q_z : \partial\Omega \to \mathbb{R}$. By Lemma 15.3 this system has a unique solution. Then evaluate

$$\begin{aligned} \tau(z) &= \int_{\partial\Omega} q_z K_z, \\ \nabla\tau(z) &= -2 \int_{\partial\Omega} q_z \nabla K_z. \end{aligned}$$

3. For $\zeta = z_1, \ldots, z_{p-1}, z_{p+1}, \ldots, z_P$ evaluate

$$\begin{aligned} G_z(\zeta) &= K_z(\zeta) - \int_{\partial\Omega} q_z K_\zeta, \\ \nabla G_z(\zeta) &= \nabla K_z(\zeta) + \int_{\partial\Omega} q_z \nabla K_\zeta. \end{aligned}$$

4. Compute $\dot{z}_p$ according to the equation of motion (15.6) and use any standard method for ordinary differential equations to update the vortex positions. In the case of a single vortex we only need to follow a level line of the hydrodynamic Robin function. Since we know its gradient this can be done at a uniform accuracy of order $O(dt^2)$.

Note that (15.11) has some similarities with the Fredholm integral equation of the second kind used for the double layer representation. In particular the kernel changes sign at the singularity. Thus it would be interesting to know whether the condition number of (15.11) also remains bounded as the mesh size tends to 0.

# Bibliography


[1] ACKER, A. An extremal problem involving current flow through distributed resistance. *SIAM J. Math. Anal. 12*, 2 (1981), 169–172.

[2] ACKER, A. On the qualitative theory of parametrized families of free boundaries. *J. Reine Angew. Math. 393* (1989), 134–167.

[3] ACKER, A., AND MEYER, R. A free boundary problem for the $p$-Laplacian: uniqueness, convexity, and successive approximation of solutions. *Electron. J. Differential Equations 8* (1995), 20.

[4] ADIMURTHI. Existence of positive solutions of the semilinear Dirichlet problem with critical growth for the $n$-Laplacian. *Ann. Scuola Norm. Sup. Pisa Cl. Sci. (4) 17*, 3 (1990), 393–413.

[5] ADIMURTHI, PACELLA, F., AND YADAVA, S. L. Interaction between the geometry of the boundary and positive solutions of a semilinear Neumann problem with critical nonlinearity. *J. Funct. Anal. 113*, 2 (1993), 318–350.

[6] AGUILERA, N., ALT, H. W., AND CAFFARELLI, L. A. An optimization problem with volume constraint. *SIAM J. Control Optim. 24*, 2 (1986), 191–198.

[7] ALT, H. W., AND CAFFARELLI, L. A. Existence and regularity for a minimum problem with free boundary. *J. Reine Angew. Math. 325* (1981), 105–144.

[8] AREF, H. Integrable, chaotic, and turbulent vortex motion in two-dimensional flows. *Annual review of fluid mechanics, Annual Reviews, Palo Alto, Calif. 15* (1983), 345–389.

[9] AREF, H., AND POMPHREY, N. Integrable and chaotic motions of four vortices. i. the case of identical vortices. *Proc. Roy. Soc. London Ser. A 380*, 1779 (1982), 359–387.

[10] AUBIN, T. Équations différentielles non linéaires et problème de Yamabe concernant la courbure scalaire. *J. Math. Pures Appl. (9) 55*, 3 (1976), 269–296.

[11] AUBIN, T. Prescribed scalar curvature and the method of isometry-concentration. In *Partial differential equations of elliptic type (Cortona, 1992)*, Sympos. Math., XXXV. Cambridge Univ. Press, Cambridge, 1994, pp. 37–45.

[12] BAERNSTEIN, A. I. A unified approach to symmetrization. In *Partial differential equations of elliptic type (Cortona, 1992)*, Sympos. Math., XXXV. Cambridge Univ. Press, Cambridge, 1994, pp. 47–91.

[13] BAHRI, A. *Critical points at infinity in some variational problems*, vol. 182 of *Pitman Research Notes in Mathematics Series.* Longman Scientific & Technical, Harlow, 1989.

[14] BAHRI, A., LI, Y., AND REY, O. On a variational problem with lack of compactness: the topological effect of the critical points at infinity. *Calc. Var. Partial Differential Equations 3*, 1 (1995), 67–93.

[15] BANDLE, C., BRILLARD, A., AND FLUCHER, M. Green's function, harmonic transplantation, and best Sobolev constant in spaces of constant curvature. *Trans. Amer. Math. Soc. 350*, 3 (1998), 1103–1128.

[16] BANDLE, C., AND ESSÉN, M. On the solutions of quasilinear elliptic problems with boundary blow-up. In *Partial differential equations of elliptic type (Cortona, 1992)*, Sympos. Math., XXXV. Cambridge Univ. Press, Cambridge, 1994, pp. 93–111.

[17] BANDLE, C., AND FLUCHER, M. Harmonic radius and concentration of energy; hyperbolic radius and Liouville's equations $\Delta U = e^U$ and $\Delta U = U^{(n+2)/(n-2)}$. *SIAM Rev. 38*, 2 (1996), 191–238.

[18] BANDLE, C., AND MARCUS, M. The positive radial solutions of a class of semilinear elliptic equations. *J. Reine Angew. Math. 401* (1989), 25–59.

[19] BANDLE, C., AND PELETIER, L. Best Sobolev constants and Emden equations on $S^3$. *submitted.*

[20] BANDLE, C., AND PORRU, G. Asymptotic behaviour and convexity of large solutions to nonlinear equations. In *Inequalities and applications*, vol. 3 of *World Sci. Ser. Appl. Anal.* World Sci. Publishing, River Edge, NJ, 1994, pp. 59–71.

[21] BETHUEL, F., BRÉZIS, H., AND HÉLEIN, F. *Ginzburg-Landau vortices*, vol. 13 of *Progress in Nonlinear Differential Equations and their Applications.* Birkhäuser Boston Inc., Boston, MA, 1994.

[22] BEURLING, A. On free-boundary problems for the laplace equation. *Sem. on analytic functions, Inst. for Advanced Study Princeton 1* (1957), 248–263.

[23] BIANCHI, G., AND EGNELL, H. An ODE approach to the equation $\Delta u + Ku^{(n+2)/(n-2)} = 0$, in $\mathbb{R}^n$. *Math. Z. 210*, 1 (1992), 137–166.

[24] BIANCHI, G., AND EGNELL, H. A variational approach to the equation $\Delta u + Ku^{(n+2)/(n-2)} = 0$ in $\mathbb{R}^n$. *Arch. Rational Mech. Anal. 122*, 2 (1993), 159–182.

[25] BRÉZIS, H. Some variational problems with lack of compactness. In *Nonlinear functional analysis and its applications, Part 1 (Berkeley, Calif., 1983)*, vol. 45 of *Proc. Sympos. Pure Math.* Amer. Math. Soc., Providence, R.I., 1986, pp. 165–201.

[26] BRÉZIS, H., AND LIEB, E. A relation between pointwise convergence of functions and convergence of functionals. *Proc. Amer. Math. Soc. 88*, 3 (1983), 486–490.

[27] BRÉZIS, H., AND NIRENBERG, L. Positive solutions of nonlinear elliptic equations involving critical Sobolev exponents. *Comm. Pure Appl. Math. 36*, 4 (1983), 437–477.

[28] BRÉZIS, H., AND PELETIER, L. A. Asymptotics for elliptic equations involving critical growth. In *Partial differential equations and the calculus of variations, Vol. I*, vol. 1 of *Progr. Nonlinear Differential Equations Appl.* Birkhäuser Boston, Boston, MA, 1989, pp. 149–192.

[29] BROTHERS, J. E., AND ZIEMER, W. P. Minimal rearrangements of Sobolev functions. *J. Reine Angew. Math. 384* (1988), 153–179.

[30] CAFFARELLI, L. A., AND FRIEDMAN, A. Asymptotic estimates for the plasma problem. *Duke Math. J. 47*, 3 (1980), 705–742.

[31] CARLEMAN, T. Über ein Minimumproblem in der mathematischen Physik. *Math. Z. 1* (1918), 208–212.

[32] CARLESON, L., AND CHANG, S.-Y. A. On the existence of an extremal function for an inequality of J. Moser. *Bull. Sci. Math. (2) 110*, 2 (1986), 113–127.

[33] CHEN, C. C., AND LIN, C.-S. Uniqueness of the ground state solutions of $\Delta u + f(u) = 0$ in $\mathbb{R}^n$; $n \geq 3$. *Comm. Partial Differential Equations 16*, 8-9 (1991), 1549–1572.

[34] COLESANTI, A. A variational problem for harmonic functions in ring-shaped domains with partially free boundary. *SIAM J. Math. Anal. 25*, 4 (1994), 1122–1127.

[35] CRANK, J. *Free and moving boundary problems.* The Clarendon Press Oxford University Press, New York, 1987.

[36] DEL PINO, M., AND FELMER, P. L. Local mountain passes for semilinear elliptic problems in unbounded domains. *Calc. Var. Partial Differential Equations 4*, 2 (1996), 121–137.

[37] DEL PINO, M., AND FELMER, P. L. Semi-classical states for nonlinear Schrödinger equations. *J. Funct. Anal. 149*, 1 (1997), 245–265.

[38] DELLACHERIE, C., AND MEYER, P.-A. *Probabilities and potential. C*, vol. 151 of *North-Holland Mathematics Studies.* North-Holland Publishing Co., Amsterdam, 1988. Potential theory for discrete and continuous semigroups, Translated from the French by J. Norris.

[39] DING, W. Y., AND NI, W. M. On the elliptic equation $\Delta u + Ku^{(n+2)/(n-2)} = 0$ and related topics. *Duke Math. J. 52*, 2 (1985), 485–506.

[40] EGNELL, H. Existence and nonexistence results for $m$-Laplace equations involving critical Sobolev exponents. *Arch. Rational Mech. Anal. 104*, 1 (1988), 57–77.

[41] EGNELL, H. Semilinear elliptic equations involving critical Sobolev exponents. *Arch. Rational Mech. Anal. 104*, 1 (1988), 27–56.

[42] EGNELL, H. Elliptic boundary value problems with singular coefficients and critical nonlinearities. *Indiana Univ. Math. J. 38*, 2 (1989), 235–251.

[43] EGNELL, H. Existence results for some quasilinear elliptic equations. In *Variational methods (Paris, 1988)*, vol. 4 of *Progr. Nonlinear Differential Equations Appl.* Birkhäuser Boston, Boston, MA, 1990, pp. 61–76.

[44] EGNELL, H. Asymptotic results for finite energy solutions of semilinear elliptic equations. *J. Differential Equations 98*, 1 (1992), 34–56.

[45] ELLIOTT, C. M., AND FRIEDMAN, A. The contact set of a rigid body partially supported by a membrane. *Nonlinear Anal. 10*, 3 (1986), 251–276.

[46] EVANS, L. C. *Weak convergence methods for nonlinear partial differential equations*, vol. 74 of *CBMS Regional Conference Series in Mathematics.* Published for the Conference Board of the Mathematical Sciences, Washington, DC, 1990.

[47] FASANO, A. Some free boundary problems with industrial applications. In *Shape optimization and free boundaries (Montreal, PQ, 1990)*, vol. 380 of *NATO Adv. Sci. Inst. Ser. C Math. Phys. Sci.* Kluwer Acad. Publ., Dordrecht, 1992, pp. 113–142.

[48] FLUCHER, M. Extremal functions for the Trudinger-Moser inequality in 2 dimensions. *Comment. Math. Helv. 67*, 3 (1992), 471–497.

[49] FLUCHER, M. An asymptotic formula for the minimal capacity among sets of equal area. *Calc. Var. Partial Differential Equations 1*, 1 (1993), 71–86.

[50] FLUCHER, M. Approximation of Dirichlet eigenvalues on domains with small holes. *J. Math. Anal. Appl. 193*, 1 (1995), 169–199.

[51] FLUCHER, M., GARRONI, A., AND MÜLLER, S. Concentration of low energy extremals: Identification of concentration points. *Draft.*

[52] FLUCHER, M., AND GUSTAFSSON, B. Vortex motion in two-dimensional hydrodynamics. *Royal Inst. Techn. Stockholm TRITA-MAT-97-MA-02* (1997).

[53] FLUCHER, M., AND KOOP, A. Numerical solution of a singularly perturbed eigenvalue problem. *University of Bonn, SFB 256 No. 309* (1993).

[54] FLUCHER, M., AND MÜLLER, S. Radial symmetry and decay rate of variational ground states in the zero mass case. *SIAM J. Math. Anal. 29*, 3 (1998), 712–719 (electronic).

[55] FLUCHER, M., AND MÜLLER, S. Concentration of low energy extremals. *Ann. Inst. H. Poincaré Anal. Non Linéaire 16*, 3 (1999), 269–298.

[56] FLUCHER, M., AND RUMPF, M. Bernoulli's free-boundary problem, qualitative theory and numerical approximation. *J. Reine Angew. Math. 486* (1997), 165–204.

[57] FLUCHER, M., AND STRUWE, M. Concentration with conformal energy. *Draft.*

[58] FLUCHER, M., AND WEI, J. Semilinear Dirichlet problem with nearly critical exponent, asymptotic location of hot spots. *Manuscripta Math. 94*, 3 (1997), 337–346.

[59] FLUCHER, M., AND WEI, J. Asymptotic shape and location of small cores in elliptic free-boundary problems. *Math. Z. 228*, (1998), 683–703.

[60] FRANCHI, B., LANCONELLI, E., AND SERRIN, J. Existence and uniqueness of nonnegative solutions of quasilinear equations in $\mathbb{R}^n$. *Adv. Math. 118*, 2 (1996), 177–243.

[61] FRIEDMAN, A. Free boundary problem in fluid dynamics. *Astérisque 118* (1984), 55–67. Variational methods for equilibrium problems of fluids (Trento, 1983).

[62] FRIEDRICHS, K. O. Über ein Minimumproblem für Potentialströmungen mit freiem Rand. *Math. Ann. 109* (1934), 60–82.

[63] GARABEDIAN, P. R. *Partial differential equations*, second ed. Chelsea Publishing Co., New York, 1986.

[64] GIDAS, B., NI, W. M., AND NIRENBERG, L. Symmetry and related properties via the maximum principle. *Comm. Math. Phys. 68*, 3 (1979), 209–243.

[65] GIDAS, B., AND SPRUCK, J. A priori bounds for positive solutions of nonlinear elliptic equations. *Comm. Partial Differential Equations 6*, 8 (1981), 883–901.

[66] GILBARG, D., AND TRUDINGER, N. S. *Elliptic partial differential equations of second order*, second ed., vol. 224 of *Grundlehren der Mathematischen Wissenschaften [Fundamental Principles of Mathematical Sciences]*. Springer-Verlag, Berlin, 1983.

[67] HAN, Z.-C. Asymptotic approach to singular solutions for nonlinear elliptic equations involving critical Sobolev exponent. *Ann. Inst. H. Poincaré Anal. Non Linéaire 8*, 2 (1991), 159–174.

[68] HERSCH, J. Transplantation harmonique, transplantation par modules, et théorèmes isopérimétriques. *Comment. Math. Helv. 44* (1969), 354–366.

[69] JOHNSON, C. *Numerical solution of partial differential equations by the finite element method.* Cambridge University Press, Cambridge, 1987.

[70] KAPER, H. G., KWONG, M. K., AND LI, Y. Symmetry results for reaction-diffusion equations. *Differential Integral Equations 6*, 5 (1993), 1045–1056.

[71] KAWOHL, B. *Rearrangements and convexity of level sets in PDE*, vol. 1150 of *Lecture Notes in Mathematics.* Springer-Verlag, Berlin, 1985.

[72] KAWOHL, B. A remark on N. korevaar's concavity maximum principle and on the asymptotic uniqueness of solutions to the plasma problem. *Math. Methods Appl. Sci. 8*, 1 (1986), 93–101.

[73] KICHENASSAMY, S., AND VÉRON, L. Singular solutions of the $p$-Laplace equation. *Math. Ann. 275*, 4 (1986), 599–615.

[74] KIMURA, Y. Chaos and collapse of a system of point vortices. *Fluid Dynamics Research 3* (1988), 98–104.

[75] KIRCHHOFF, G. *Vorlesungen über mathematische Physik.* 1876.

[76] KWONG, M. K. Uniqueness of positive solutions of $\Delta u - u + u^p = 0$ in $\mathbb{R}^n$. *Arch. Rational Mech. Anal. 105*, 3 (1989), 243–266.

[77] KWONG, M. K., AND LI, Y. Uniqueness of radial solutions of semilinear elliptic equations. *Trans. Amer. Math. Soc. 333*, 1 (1992), 339–363.

[78] KWONG, M. K., AND ZHANG, L. Q. Uniqueness of the positive solution of $\Delta u + f(u) = 0$ in an annulus. *Differential Integral Equations 4*, 3 (1991), 583–599.

[79] LACEY, A. A., AND SHILLOR, M. Electrochemical and electro-discharge machining with a threshold current. *IMA J. Appl. Math. 39*, 2 (1987), 121–142.

[80] LAZER, A. C., AND MCKENNA, P. J. Asymptotic behavior of solutions of boundary blowup problems. *Differential Integral Equations 7*, 3-4 (1994), 1001–1019.

[81] LEVITSKIĬ, B. E. Reduced $p$-modulus and the interior $p$-harmonic radius. *Dokl. Akad. Nauk SSSR 316*, 4 (1991), 812–815.

[82] LI, Y. Asymptotic behavior of positive solutions of equation $\Delta u + K(x)u^p = 0$ in $\mathbb{R}^n$. *J. Differential Equations 95*, 2 (1992), 304–330.

[83] LI, Y. Symmetry properties of finite total mass solutions of Matukuma equation. In *Nonlinear diffusion equations and their equilibrium states, 3 (Gregynog, 1989)*, vol. 7 of *Progr. Nonlinear Differential Equations Appl.* Birkhäuser Boston, Boston, MA, 1992, pp. 375–389.

[84] LI, Y., AND NI, W.-M. On the asymptotic behavior and radial symmetry of positive solutions of semilinear elliptic equations in $\mathbb{R}^n$. I. Asymptotic behavior. *Arch. Rational Mech. Anal. 118*, 3 (1992), 195–222.

[85] LI, Y., AND NI, W.-M. On the asymptotic behavior and radial symmetry of positive solutions of semilinear elliptic equations in $\mathbb{R}^n$. II. Radial symmetry. *Arch. Rational Mech. Anal. 118*, 3 (1992), 223–243.

[86] LI, Y., AND NI, W.-M. Radial symmetry of positive solutions of nonlinear elliptic equations in $\mathbb{R}^n$. *Comm. Partial Differential Equations 18*, 5-6 (1993), 1043–1054.

[87] LIN, C. C. *On the Motion of Vortices in Two Dimensions.* University of Toronto Press, Toronto, Ont., 1943. University of Toronto Studies, Applied Mathematics Series, no. 5.

[88] LIN, K.-C. Extremal functions for Moser's inequality. *Trans. Amer. Math. Soc. 348*, 7 (1996), 2663–2671.

[89] LIONS, P.-L. The concentration-compactness principle in the calculus of variations. The locally compact case. I. *Ann. Inst. H. Poincaré Anal. Non Linéaire 1*, 2 (1984), 109–145.

[90] LIONS, P.-L. The concentration-compactness principle in the calculus of variations. The locally compact case. II. *Ann. Inst. H. Poincaré Anal. Non Linéaire 1*, 4 (1984), 223–283.

[91] LIONS, P.-L. The concentration-compactness principle in the calculus of variations. The limit case. I. *Rev. Mat. Iberoamericana 1*, 1 (1985), 145–201.

[92] LIOUVILLE, J. Sur l'équation aux différences partielles $\partial^2 \log \lambda / \partial u \partial v \pm \lambda / 2a^2 = 0$. *J. de Math. 18*, 1 (1853), 71–72.

[93] MARCHIORO, C., AND PULVIRENTI, M. *Mathematical theory of incompressible nonviscous fluids*, vol. 96 of *Applied Mathematical Sciences*. Springer-Verlag, New York, 1994.

[94] MCLEOD, J. B., AND PELETIER, L. A. Observations on Moser's inequality. *Arch. Rational Mech. Anal. 106*, 3 (1989), 261–285.

[95] MINDA, D., AND OVERHOLT, M. The minimum points of the hyperbolic metric. *Complex Variables Theory Appl. 21*, 3-4 (1993), 265–277.

[96] MORREY, CHARLES B., J. *Multiple integrals in the calculus of variations*. Springer-Verlag New York, Inc., New York, 1966. Die Grundlehren der mathematischen Wissenschaften, Band 130.

[97] MOSER, J. A new proof of De Giorgi's theorem concerning the regularity problem for elliptic differential equations. *Comm. Pure Appl. Math. 13* (1960), 457–468.

[98] MOSER, J. A sharp form of an inequality by N. Trudinger. *Indiana Univ. Math. J. 20* (1970/71), 1077–1092.

[99] NI, W.-M., AND TAKAGI, I. Locating the peaks of least-energy solutions to a semilinear Neumann problem. *Duke Math. J. 70*, 2 (1993), 247–281.

[100] NI, W.-M., AND TAKAGI, I. Spike-layers in semilinear elliptic singular perturbation problems. In *Degenerate diffusions (Minneapolis, MN, 1991)*, vol. 47 of *IMA Vol. Math. Appl.* Springer, New York, 1993, pp. 131–139.

[101] NI, W.-M., AND WEI, J. On the location and profile of spike-layer solutions to singularly perturbed semilinear Dirichlet problems. *Comm. Pure Appl. Math. 48*, 7 (1995), 731–768.

[102] POINCARÉ, H. Sur le problème de la distribution électrique. *Comptes rendus de l'Academie des sciences 104, ser. 1* (1887), 44–46.

[103] PÓLYA, G., AND SCHIFFER, M. Convexity of functionals by transplantation. *J. Analyse Math. 3* (1954), 245–346. With an appendix by Heinz Helfenstein.

[104] PÓLYA, G., AND SZEGÖ, G. *Isoperimetric Inequalities in Mathematical Physics*. Princeton University Press, Princeton, N. J., 1951. Annals of Mathematics Studies, no. 27,.

[105] REN, X., AND WEI, J. On a two-dimensional elliptic problem with large exponent in nonlinearity. *Trans. Amer. Math. Soc. 343*, 2 (1994), 749–763.

[106] REN, X., AND WEI, J. On a semilinear elliptic equation in $\mathbb{R}^2$ when the exponent approaches infinity. *J. Math. Anal. Appl. 189*, 1 (1995), 179–193.

[107] REY, O. A multiplicity result for a variational problem with lack of compactness. *Nonlinear Anal. 13*, 10 (1989), 1241–1249.

[108] REY, O. Proof of two conjectures of H. Brézis and L. A. Peletier. *Manuscripta Math. 65*, 1 (1989), 19–37.

[109] REY, O. Bifurcation from infinity in a nonlinear elliptic equation involving the limiting Sobolev exponent. *Duke Math. J. 60*, 3 (1990), 815–861.

[110] REY, O. The role of the Green's function in a nonlinear elliptic equation involving the critical Sobolev exponent. *J. Funct. Anal. 89*, 1 (1990), 1–52.

[111] REY, O. Concentration of solutions to elliptic equations with critical nonlinearity. *Ann. Inst. H. Poincaré Anal. Non Linéaire 9*, 2 (1992), 201–218.

[112] REY, O. Le rôle de la frontière dans un problème elliptique avec nonlinéarité critique et conditions au bord de Neumann. In *Séminaire sur les Équations aux Dérivées Partielles, 1993–1994*. École Polytech., Palaiseau, 1994, pp. Exp. No. XXIII, 13.

[113] REY, O. Boundary effect for an elliptic neumann problem with critical nonlinearity. *Comm. Partial Diff. Eq. 22* (1997).

[114] ROKHLIN, V. Rapid solution of integral equations of classical potential theory. *J. Comput. Phys. 60*, 2 (1985), 187–207.

[115] ROUTH, E. Some applications of conjugate functions. *Proc. London Math. Soc. 12* (1881), 73–89.

[116] SCHIFFER, M. Partial differential equations of the elliptic type. *Lecture series of the symposium on partial differential equations, University of California* (1955), 97–149.

[117] SCHOEN, R. Conformal deformation of a Riemannian metric to constant scalar curvature. *J. Differential Geom. 20*, 2 (1984), 479–495.

[118] SIMON, J. Differentiation with respect to the domain in boundary value problems. *Num. Funct. Anal. Optimiz. 2* (1980), 649–687.

[119] SOKOŁOWSKI, J., AND ZOLÉSIO, J.-P. *Introduction to shape optimization*, vol. 16 of *Springer Series in Computational Mathematics*. Springer-Verlag, Berlin, 1992. Shape sensitivity analysis.

[120] STRUWE, M. Critical points of embeddings of $H_0^{1,n}$ into Orlicz spaces. *Ann. Inst. H. Poincaré Anal. Non Linéaire 5*, 5 (1988), 425–464.

[121] STRUWE, M. *Variational methods*, second ed., vol. 34 of *Ergebnisse der Mathematik und ihrer Grenzgebiete (3) [Results in Mathematics and Related Areas (3)]*. Springer-Verlag, Berlin, 1996. Applications to nonlinear partial differential equations and Hamiltonian systems.

[122] SZEGÖ, G. Über einige Extremalaufgaben der Potentialtheorie. *Math. Z. 31* (1930), 583–593.

[123] TALENTI, G. Elliptic equations and rearrangements. *Ann. Scuola Norm. Sup. Pisa Cl. Sci. (4) 3*, 4 (1976), 697–718.

[124] TEICHMÜLLER, O. Untersuchungen über konforme und quasikonforme abbildung. *Deutsche Math. 3* (1938), 621–678.

[125] TEMAM, R. Numerical problems in plasma physics. In *Numerical methods for partial differential equations*, vol. 42 of *Proc. Adv. Sem., Math. Res. Center, Univ. Wisconsin, Madison, Wis., 1978.* Academic Press, New York-London, 1979, pp. 307–327.

[126] TEPPER, D. E. A mathematical model for a wake. *Michigan Math. J. 31*, 2 (1984), 161–165.

[127] TOLKSDORF, P. On the Dirichlet problem for quasilinear equations in domains with conical boundary points. *Comm. Partial Differential Equations 8*, 7 (1983), 773–817.

[128] TREFETHEN, L. N. Erratum: "Numerical computation of the Schwarz-Christoffel transformation". *SIAM J. Sci. Statist. Comput. 1*, 2 (1980), 302.

[129] TREFETHEN, L. N. Numerical computation of the Schwarz-christoffel transformation. *SIAM J. Sci. Statist. Comput. 1*, 1 (1980), 82–102.

[130] TRUDINGER, N. S. On imbeddings into Orlicz spaces and some applications. *J. Math. Mech. 17* (1967), 473–483.

[131] WLOKA, J. *Partial differential equations.* Cambridge University Press, Cambridge, 1987. Translated from the German by C. B. Thomas and M. J. Thomas.

[132] YANAGIDA, E. Uniqueness of positive radial solutions of $\Delta u + g(r)u + h(r)u^p = 0$ in $\mathbb{R}^n$. *Arch. Rational Mech. Anal. 115*, 3 (1991), 257–274.

# Index

MIX
Papier aus verantwortungsvollen Quellen
Paper from responsible sources
FSC® C105338

If you have any concerns about our products,
you can contact us on
**ProductSafety@springernature.com**

In case Publisher is established outside the EU,
the EU authorized representative is:
**Springer Nature Customer Service Center GmbH**
**Europaplatz 3, 69115 Heidelberg, Germany**

Printed by Libri Plureos GmbH
in Hamburg, Germany